PATRICK GALBRAITH was born in Edinburgh in 1993. He grew up in Scotland and studied English literature at Bristol and University College London before dropping out to become a writer and journalist. He has covered everything from naturism to nightingale conservation to racism in rural Britain and his work has appeared in *The Times*, *The Telegraph*, *The Spectator* and *Country Life*. He is commissioning editor of *The Open Art Fair Magazine* and is a columnist for *The Critic*. He lives in South London with his terrier and is currently working on an experimental production about lapwings, a bird that he has long felt deserves more time on the stage.

'A treasury of a book . . . filled with beautiful moments, amazing and sometimes rather surprising characters, and, if we could only learn from them, reasons for hope.'

John Burnside, *New Statesman*

'One of the great oral histories of British nature and the British countryside . . . Sad and honest and important and often very funny' Richard Smyth, *Review 31*

'It's a delight to jump into this slightly strange parallel world. Galbraith is such an able communicator of its weirdness, that it is a pleasure to go along for the ride' *The Times*

'A cracking read . . . Without a doubt in my mind, it's my book of the year' Mark Avery

'In terms of both scope and execution, this book is a hugely impressive achievement, and it will be fascinating to see where Galbraith goes from here.' *The Scotsman*

'An important and timely book that explores the human context of an ecological emergency. Galbraith is a thoughtful, assured and

elegant writer who brings a mature intelligence and open-minded insight to his subject.'
Oban Times

'Galbraith is an inquisitive people-watcher with an unsparingly honest eye for detail – he captures the quirkiness of many of those on the frontline of conservation . . . Entertaining . . . He takes us to some of the wildest and remotest parts of Britain and describes them with an empathy for the agricultural stories behind the landscape . . . It is a hopeful sign that he takes no side in the culture wars and seeks instead to bring everyone together and seek practical solutions to the prevention of the last song of some of our best-loved birds.'
Jamie Blackett, *Country Life*

'Galbraith's writing is beautiful . . . *In Search of One Last Song* feels like an important step in the right direction'
Stephen Rutt, *British Birds*

'The birds come to life in his fine writing'
Helen Bynum, *Literary Review*

'Galbraith combines the ability to write lyrically with a formidable grasp of his subject'
The Week

'The writing is strong, the book an impressive debut, establishing Galbraith as a quality writer.' Tim Dee, Caught by the River

'A requiem for cherished birds; an overture to another threatened species, the rural people who know Nature's ways (but done without nostalgia, because Galbraith is young and an insider). All this in a perfectly written travelogue – the best book on conservation and the countryside I have read in years.'
John Lewis-Stempel, 'Britain's finest living
nature writer' (*The Times*)

'Unexpected . . . very readable'   *The Countryman*

'Patrick Galbraith's engaging debut volume will appeal to the layman as much as to the committed naturalist, being a quirkily enjoyable journey through a slightly nether worldly version of Britain . . . The book is wonderfully free from eco-lecturing, and one of its very strengths is the way Galbraith maintains a neutral tone despite the gaggle of different voices from those he meets.'

David Profumo, *The Critic*

'I strongly recommend this book . . . [Galbraith] explor[es] a cultural landscape as much as a physical one, in which human creativity is an expression of the natural world'   Katrina Porteous

'A vivid account of lives lived alongside our most iconic and endangered birds – and a glimpse of the cultural void they would leave behind. Galbraith has a poet's eye, his descriptions are crystalline, deeply affecting, while his encounters with others are direct, and occasionally ruffling – *In Search of One Last Song* brings to mind the writing of Bruce Chatwin, in the best of ways.'

Cal Flyn, bestselling author of *Islands of Abandonment*

'What a cornucopia this is! I have had a wonderful, enriching morning reading *In Search of One Last Song*. It is a rucksack full of experience and deep knowledge, full of love for part-hidden treasure in the country beneath the country most of us know. I am left with a sense of the shocking carelessness of what we have all done, the way in which creatures need to find their niches now only in the fragmentary and abused corners we have left for them. But alongside that is the recognition of how utterly valuable the unacknowledged and unofficial knowledge is, the sheer understanding of those who have long looked after and known the country.'

Adam Nicolson, winner of the Somerset Maugham,
Wainwright, and Costa biography prizes

'*In Search of One Last Song* is steeped in loss; an evocative and beautifully-written history of the complex relationship between Britain's people and its birds.'

Jonathan Slaght, author of the award-winning
Owls of the Eastern Ice

'A triumph. Well worth a meditative read, a love song to Britain's birds and to those who love them.'

Fergus Butler-Gallie, *Church Times*

'Beautifully written and earthy'

Philip Womack, *The London Magazine*

'This is no clinical ornithological autopsy. Instead, it's a series of intimate portraits of remarkable people for whom birds represent a personal and cultural identity . . . Galbraith writes with an intimacy and fondness that connects readers to the birds through their quirky defenders.'   *The Spectator World*

'A sterling piece of journalism and a funny, warm study of people who don't often make the headlines. It's the editor's book of the year.'

Charlie Baker, *The Fence*

IN SEARCH OF
ONE LAST SONG

**BRITAIN'S DISAPPEARING BIRDS AND THE
PEOPLE TRYING TO SAVE THEM**

PATRICK GALBRAITH

**WILLIAM
COLLINS**

William Collins
An imprint of HarperCollins*Publishers*
1 London Bridge Street
London SE1 9GF

HarperCollins*Publishers*
Macken House
39/40 Mayor Street Upper
Dublin 1
D01 C9W8
Ireland

First published in Great Britain in 2022 by William Collins
This William Collins paperback edition published in 2023

1

Copyright © Patrick Galbraith 2022
Illustrations by Robert Vaughan

Patrick Galbraith asserts the moral right to be identified as the author of this
work in accordance with the Copyright, Designs and Patents Act 1988

A catalogue record for this book is available from the British Library

ISBN 978-0-00-842050-5

Set in Adobe Garamond Pro by Palimpsest Book Production Ltd,
Falkirk, Stirlingshire

Printed and bound in the UK using 100% renewable
electricity at CPI Group (UK) Ltd

This book is produced from independently certified FSC™ paper
to ensure responsible forest management.

For more information visit: www.harpercollins.co.uk/green

For Constance, for my uncles too, who I hope would have enjoyed this book very much, and for all those people trying to preserve the beauty of the world.

Contents

Preface

Below the burn: Dumfriesshire, autumn 2008

Up behind the river, at the back of the kirkyard, the rabbits lived among the dead. I wasn't ever sure if I was meant to be shooting in the grave-yard itself but I was fourteen and Walter, a sheep farmer who sat on the parish council, rang one evening to say he'd lend me his air rifle if I fancied it. The flowers left by grieving relatives were being eaten and it had been decided that something had to be done.

When the sun shone, I'd cycle over on my younger brother's bike and leave it leaning on the ivy-covered gable end of the old ruined chapel next to the pink sandstone kirk. All afternoon and often long into the evening, I'd crawl among the graves, head up into the fields, and wander in the birch wood beyond, trying to do the best I could to thin out the rabbits. There were hundreds then, but even when I learned how to use the wind and the stone dykes to get right in among them, I only ever went home with two or three. I'm sure it meant though that Walter could say yes, absolutely, the flowers being eaten was a problem he was dealing with.

On my way back home, after the humpbacked bridge and past the rowans, where the railway line once ran, I'd always stop to call

1

in on the old lady who lived down the lane. From her kitchen window, she could see the gravestones in the distance and she once told me the story of how she'd met her husband, who lay beneath one of them, just after the war, but mostly she wanted to talk about the hills beyond the kirkyard where she'd lived as a little girl. Back then, in spring, she remembered peewits and curlew coming crying across the fields and there were so many black grouse she told me nobody would have believed you if you'd said that one day they'd be all but gone. Often, the following day, I'd return with some rabbit casserole – 'Made it myself, cider and honey and mustard. You just have to put it in the microwave' – then we'd sit and talk some more. She had a black cat called Titus that would lie across her lap and I'm not sure I ever saw him awake. One morning, she showed me a map of the Shinnel Glen and told me there was a pool there where you could guddle trout. 'You rub your fingers under their bellies,' she said, in a whisper, showing me how with her small white hands, 'then when they fall into a daze, you grab them behind the gills and quickly hoick them onto the bank.'

That afternoon I cycled up there with my terrier. The sun hung orange in the quiet sky and beech mast and leaves lay drifted on the lane but the burn was on its bones and there were no fish where she said they'd be. For a couple of hours I stalked the shallows then I sat in the moss and looked down the glen, wondering if without all those birds it was really the same place at all.

Introduction

Waterloo Station, January 2020

Chris Packham and I had been in touch for a while and we thought that maybe we could do something together. I knew plenty of people who seemed to hate him, and as the editor of a prominent fieldsports magazine he knew plenty of people who felt the same way about me. It was a cold Monday afternoon and I don't think it had actually rained all day but the streets were greasy and damp.

I checked the email again as my train pulled in. His assistant had sent it four weeks previously and David – whoever David was – had apparently put the date in his calendar: 'There is a restaurant on the concourse at Waterloo Station, the Natural Kitchen, it might work as a place to meet.' Chris was sitting with his glasses on the tip of his nose scrolling through his phone as I approached. He looked older than he does on telly and he was pushing the corner of his lip up with his tongue as though something was frustrating him, but when he noticed me lingering he looked round and smiled. 'Patrick,' he said, standing up. 'I'm sorry it's not very . . .' He gesticulated apologetically and laughed, then said he'd eaten, 'but you have whatever you like. Order whatever.'

Chris Packham: Forever Punk had just aired and I told him how much I'd enjoyed it and the way the three episodes sort of captured what it was to be a teenager and just that whole moment. 'You know,' he replied, 'making that was just terrific.' As I picked at my fried courgette, we talked of fishing. We'd both done a bit when we were young. He told me he wished he wrote more and I told him I wish I fished more. 'Too busy,' he said with a shrug. 'Yeah, life does get like that,' I replied. When I was finished, he ordered tea, and when it came, the conversation turned to what we could maybe do together to try and stem the decline of some of our disappearing birds. I didn't go for his ideas – too much, too antagonistic, too divisive at a time when I felt we needed less division – and he didn't go for mine. Too slow, too plodding. 'Sometimes, Patrick, you've got to make a bit of noise.'

Below us, the station was starting to fill up with agitated commuters trying to get back home. Somewhere down the line, there'd been an incident, and emergency services were responding. Chris's train wasn't going to be for another half hour so we ordered more tea and talked on. We spoke of some of the birds we loved and how little time some of them have left. In a way I hadn't really anticipated, he listened intently, and I tried to listen just as intently to him. At just after five we wandered over to the escalator and descended into the crowd, a mass of grey faces inches apart and bodies pressed together. That morning, almost 6,000 miles away in Thailand, a sixty-one-year-old traveller in a Bangkok hospital was confirmed to be carrying a virus that hadn't previously been seen outside China.

We stood for a moment looking up at the departure boards. 'It'd be a good night for ducks,' I said to Chris. He checked his tickets then turned to look at me. 'Almost a full moon,' he replied, then he swung his bag onto his shoulder and disappeared through the

4

barriers. Outside, a storm was rising and beyond the glass canopy, lights shone down, an orange glow on oily black tracks.

The crowd grew and grew and for 20 minutes I waited for my train and then when it came, it was mostly empty and when it left, it seemed almost nobody had got on. At Vauxhall, I got off and crossed the river. Down at the water's edge, wings outstretched, two cormorants were silhouetted against the mud. Chris and I had talked and talked. Few people work harder than he does to try and make people understand the plight of some of the birds we're set to lose, and over the years, as a journalist, I've tried to do the same. I felt very aware, though, that while we were talking about what we could do or might do, there were people across the country already doing, people who have quietly devoted everything to saving Britain's birds and the places they need.

By the time I got to Pimlico, I'd decided I needed to go and meet those people, the animal rights activists and the gamekeepers, regenerative farmers and scientists. I understood that some of them mistrusted each other deeply and recognised that some of them would mistrust me, but I also felt that from different directions we were all moving towards the same end. I knew a little of the ways in which crafts like hedgelaying, coppicing and reed cutting maintained vital habitats but I wanted to actually understand. It dawned on me too, in the toilets of The Gallery pub on Lupus Street, that if I didn't hear a nightingale, a turtle dove or a capercaillie soon, I probably never would. For a long time, I'd been thinking about the presence of birds in our culture, the way they animate soundscapes, and what they represent to those who live alongside them. I wanted to know what we're really set to lose when our birds have gone and I decided then to set out in search of one last song.

21.24 grams

The thrushes sing as the sun is going,
And the finches whistle in ones and pairs,
And as it gets dark loud nightingales
In bushes
Pipe, as they can when April wears,
As if all Time were theirs.

Thomas Hardy, 'Proud Songsters',
1928, a year before he died

On Tower Bridge, an old man and his bulldog wander in the sun and three women run, one behind the other, but the tourists are gone. It is mid-March 2020 and the warmest and wettest winter on record is finally drying out into spring. At two o'clock, as the latest news from Italy hits the headlines, I hang a right onto The Highway. In the nineteenth century, the ancient Roman road, running from Whitechapel to Limehouse, was one of London's saltiest streets. As a child, before I snapped the arm off my father's record player, I used to listen to a recording of Ronnie Drew, the Irish folk musician, singing a scorched warning to sailors to stay away from The Highway's whorehouses.

Drew doesn't say anything about it, but for seamen in search of something to do, who were happy to forgo the girls, there was Jamrach's Exotic Animal Emporium at number 179. Doubtless, selling exotic animals comes with risks, and in 1857 Charles Jamrach was ordered to pay £60 to an eight-year-old boy who was carried off down the road in the jaws of an escaped Bengal tiger. The East End child had never seen such a cat but assumed, like The Highway's sooty strays, it was in want of a rub behind the ears. Jamrach's stock was varied and in *Good Words*, an eclectic fireside read pitched at the pious masses, an article was published noting that, along with the bears and a bird that looked 'so much like an old gentleman', there was an 'Asiatic deer' for £15.

Grinding down The Highway, hanging on to second gear for too long while guessing, pointlessly, where the Emporium stood, I wonder if it might have been a muntjac, the same species I'm going in search of in Suffolk. If the mysterious deer was a muntjac, it wouldn't have been the first of its kind to be unloaded in London. The city's Zoological Society had a few as early as 1838 but it is Herbrand Russell, 11th Duke of Bedford, who is generally held responsible for the muntjac's first flight. Russell's wife, Mary, a student of jujitsu and aviation, kept detailed records of her husband's animals. According to her diaries, the Duke got hold of a number of Indian muntjac in around 1890, which became aggressive on being released at his home, Woburn Abbey, and after killing a beloved dog, were sentenced to face the firing squad. Following the incident he acquired some of the Reeves variety. The releases went more peacefully and by 1905 twenty-four had been unloaded at Woburn, of which a small number had found gaps in walls and fences and were making their way in the wild.

A century later, it's no longer known how many there are in the UK. The fuzzy consensus is that we lost sight of the figures a couple of decades ago, when the population passed 50,000. At about the same time, it started to be realised that as the muntjac population was rising, nightingale numbers were tumbling. The idea that the abundant present population of muntjac all started with a few bold escapees is well worn. In truth, though, Hastings Russell, the 11th Duke's son – a fascist who fed his parrot roast beef – was the architect of the species' success. In the mid-twentieth century he carried on his father's work by releasing mixed groups of muntjac into areas of suitable habitat, which eventually connected and multiplied.

At just 20 inches tall, muntjac rip through the undergrowth and browse down embryonic plants almost as soon as they push up out of the earth. If uneaten, these seedlings become perfect for the nightingale by growing up into dense scrub. Observing their chosen habitat, medieval mythmakers believed nightingales were terrified of hungry snakes that slithered across the woodland floor and weaved tales in which the birds pressed their breasts to thorns to avoid falling fatefully asleep. In the seventeenth century, Sir Thomas Browne, the melancholy Norfolk naturalist, poured cold water on the sleepless myth. It wasn't, he believed, that nightingales pierced their hearts with thorns but that they used thorns to save their hearts from any passing predator. Of all the muntjac Hastings Russell released across England, before going into the bushes and shooting himself in the head, suicide not confirmed, the only group that didn't succeed was in Kent, one of the few places nightingales can still be heard singing in the woods at night.

At Canary Wharf, where the giant neon-signed Barclay's tower looms over Billingsgate fish market, I kill the radio and watch London fall away in the rearview mirror.

The track to Jim Allen's cottage is full of holes and by the time we make it out onto the lane, the coffee he passed me before we walked out the door is all over my thighs. As we wind our way north, through the outskirts of Ipswich towards second-homeowner Suffolk, the pavements are still busy. Children are riding their bicycles home from school and mothers push prams in the park.

As a marine, Jim 'did Iraq and Afghan in the heydays' and then served off the coast of Somalia but he didn't enjoy his time at sea. Streets roll out into hungry mid-March fields, still far too sodden to drill, and Jim speaks pitiably about the pirates they were tackling. 'Just trawlermen,' he reckons, who'd been recruited by powerful criminal gangs after being bullied off their traditional fishing grounds by large international operators. 'Nowadays, I'm a nanny really,' he tells me contentedly, as we turn down a single-track lane with the remnants of handsome hedges, grown wispy and thin, running along the verges. Jim seems to sense my confusion and explains he works in private security for wealthy families. Usually, he continues, they come to realise there's no great threat to their lives and he's deployed to take a new puppy out for a pee or to pick the children up from school. I like the thought of big Jim lumbering down the pavement with a recorder case under his arm, listening patiently to a little boy telling him how he intends to breathe life into Joseph in the Christmas nativity play.

While his job demands odd hours it leaves him a lot of time for deer management, which he's been involved in since he was a boy. Jim shoots around a hundred animals a year for the Suffolk Wildlife Trust but he also works across East Anglia for private landowners. Back when he was young, he recalls that muntjac 'were still some-thing of a novelty', but that has changed completely. When he's on Trust land, Jim often meets dog walkers who ask him what he's shooting. 'When I say I'm after muntjac, they're usually pleased.' Jim tells me, three times, he's never accepted money to take someone hunting and he believes the commercialisation of deer control contributes to their overabundance. 'You've got to be pretty ruthless,' he insists. 'We're never going to eradicate muntjac in this country. We can only stay on top of them, but people are leaving animals because they think that'll grow into a nice buck.' A nice buck, according to Jim, can often fetch up to 1,000 pounds from European hunters who like the idea of having the animal's sharp antlers on their wall. This preference for targeting large male deer, which would doubtless keep Freud up at night, has been cited as part of the problem. Any effective culling programme needs to account for large numbers of females, but there is limited interest. Jim also believes there is truth in tales of hunting agents carting muntjac around the country to start new populations that they hope to be able to monetise. 'After all, they're in Ireland now,' he says with a shrug, 'and they've hardly swum there.'

We drive off the road into the trees and Jim brings his truck to a stop. He climbs out, walks round to the boot, and then reappears with a rifle slung over his shoulder and a finger pressed to his lips. I close the passenger door as quietly as possible and follow on. Jim has slipped off ahead of me under a fence and into an open field. When I catch up he is leaning on an old post, looking out among the scraggy crops. A hare, motionless too, stares back at him, wide-eyed, awaiting our next move. Then it turns and bounds away up a tramline and out of sight. Thirty yards on, Jim cuts in next to a row of beeches and disappears into the trees. This time, when I draw level, he has relaxed. He tells me there are sometimes muntjac browsing on the ride that stretches out through the 20-acre wood, but there's nothing beyond the three pigeons clattering away out of the firs.

Every wood has a story. The one we're in is known as Gold Medal Wood but Jim shakes his head absently when I ask why. I later learn that in November 1793, Samuel Kilderbee, the town clerk of Ipswich, planted 10 acres of acorns on the site. Kilderbee recorded that barbed 'ashen keys, white-thorn berries, and whin seeds were sown with a design to protect and nurse up the produce of the acorns, and to prevent their being destroyed by hares or rabbits.' Despite Kilderbee's concerns that the acorns were not 'of the best quality', the protective measures were a success and he was awarded a gold medal by the Society for the Encouragement of Arts, Manufacturers and Commerce.

Deeper into the woods, past an old timber barn, which has escaped being transformed into someone's slice of long-weekend wonder, a woodcock flits out of a patch of brambles and dances away, silent-winged, through the trees. A month later, when thousands are dying and I'm reading about Kilderbee, I notice he refers to the land at Gold Medal Wood as 'woodcock soil' and I think of the bird, somewhere over the sea, migrating east to its northern breeding grounds.

Jim has been managing deer in the area for five years and believes he's now just about on top of them, but he sometimes feels as though as fast as he can shoot muntjac, they return. His gaze is fixed on the ground in front of us and every couple of paces he points to hoof prints in the mud and tells me how fresh they are. 'Those are from fallow,' he explains, pushing the end of his stick towards two deep gouges about 2 inches long, 'no more than a few hours old, and those are muntie slots, could be yesterday.' The difference is stark, with the muntjac hooves leaving a delicate 1-inch tread. Then he stops and points at a gnarled tangle of hawthorn and bramble. 'See that – there was hardly any of that when I started.' His voice chimes with joy. 'Deer had eaten the entire understorey down to the ground but it's coming right back'. When we get to the high seat that Jim has built among the trees, he gestures for me to go first and I clamber up the ladder, my feet slipping on the mossy rungs. Sitting down, I realise we are in the middle of the wood, where two rides crisscross, and looking down along them I can see right out across the fields.

Another big problem, Jim whispers, after he has climbed up and sat down next to me, is that deer fetch such low prices from game dealers there is no incentive to shoot them. The issue is that it is much easier and cheaper for supermarkets to utilise a steady supply of farmed venison from New Zealand than to bother with meat that's been harvested in the UK – 'It's the Kiwi boys that have flooded us.' In years past, Jim butchered all the deer he shot and sold the venison at a local farmers' market on a Saturday morning. He loved setting off just after dawn, he tells me, 'truck full of all sorts', but because of work he kept letting them down.

At a quarter to five, the watery yellow sky fades to grey and the coolness of evening steals in. I tilt my head back and purse my lips. I want to see my breath hang in the air, to make me feel it's okay that I'm shivering and my fingers are going numb. Up above us, wood pigeons slice through the wind and flutter down into the warmth of the ivy-covered oaks to roost for the night. Jim hands me the rifle. Muntjac are busiest at dawn and dusk, and if we are to see one it will be in the next 45 minutes before darkness falls.

Jim is trying to recall the last time he heard a nightingale in Suffolk when he stops, passes me the thermal imager he's been peering through and tells me there's a muntjac right down at the end of the ride. Bringing it up to my eye, I can see a hunched illu-mined outline browsing among the daffodils. In the conservative world of deer management, the device I'm holding was treated with suspicion when they first became available a decade ago, but they are now deemed essential by anybody attempting to control numbers in a meaningful way.

For five minutes we sit, saying nothing, and I peer down the ride, straining my eyes to try and see the deer. Jim is looking through the imager fixedly but then puts it down, sucks the air between his teeth, and tells me that the buck's disappeared. 'I don't often bother trying to call them but it's worth a go.' He then pulls a little rubber ball from his pocket and as his big hand closes round it, a plaintive squeal fills the silence. The sound is meant to appeal to the muntjac's protective nature by imitating a distressed fawn. Before I spot it, Jim's already given the command – 'As soon as it's out in the open, take the shot.' Then I see it too, trotting dutifully between the trees before emerging into the last of the light. I bring the rifle up and place the cross hairs just behind its front leg, tracking it as it moves. Red deer, set against a snowy hillside, have a certain majesty, and roe are bucolic, but muntjac – with their dainty gait and curious doe-eyed gaze – are the prettiest of the deer. My mouth goes dry and I think about the way the whims of wealthy Victorians have led to the impoverishment of woodland soundscapes and the creature in front of me becoming 'vermin'. Then it stops, turns to look up at us and my mind is blank. I squeeze the cold trigger: a hollow click and a jay somewhere above us cries out as the buck tears through the trees. 'It happens to all of us,' Jim says with a smile and I don't have anything to say at all, except I was sure I'd put a round in the chamber.

In 2010, some years after he was arrested for stirring up racial hatred following a speech made at a country fair, Robin Page, the self-styled vocal yokel, wrote a piece for the *Daily Mail* on the 'immigrant' deer. Page used the article to rewrite history, declaring that muntjac first set their hooves in England in 1900. They are, he lamented, 'taking over' and 'breeding like rabbits'. Below the line, praise for Page's 'clever use' of muntjac, as a proxy to air trampled truths, rolled in. A year later, Rory Knight Bruce, journalist and keen foxhunting man, revealed, in the same paper, that there are now an extraordinary two million muntjac in Britain, making him possibly the only and certainly the least-qualified person to commit to any sort of population estimate.

Since the mid-nineties, Ross Guyton, who currently heads up a team of forestry consultants, has devoted himself to restoring English woodland. Deer, he tells me, are the biggest barrier to success, but he believes muntjac often bear the brunt of ecologically undue hatred. He accepts that the muntjac population is the fastest growing of the five species of deer in Britain, but he thinks attitudes towards them are so obsessive that they detract from damage caused by fallow, which he sometimes sees in herds of up to 200. Fallow were first introduced by the Romans, later becoming extinct after the collapse of the Roman Empire, before being reintroduced by the Normans for hunting. While they are non-native they are considered to be naturalised. Ross thinks this has led to them being perceived as acceptable whereas muntjac, he reckons, are viewed as a 'dirty exotic', busily displacing a sonorous symbol of merrie olde England.

Thirty-five miles inland of where I sat out with Jim, lies Bradfield Woods. Since 1252, the 200-acre area has been managed to supply thatchers and toolmakers. In the 1980s, the choir of nightingales, which had sung there for as long as anybody could remember, began to steadily disappear. In response, Dr Rob Fuller of the British Trust for Ornithology conducted an experiment that saw 6 per cent of the woodland area enclosed with deer fencing. Seven male nightingales were fitted with tiny radio transmitters so that their movements could be tracked. Within a short period, the birds started to show a very clear preference for the deer-free areas, eventually spending 69 per cent of their time there. In their write-up, the team noted that the preference for the exclusion zone may well have been higher but there was finite space, so the early birds got dense tranquillity and the others had to make do in the ravaged thin lands beyond. The study is often used to highlight the harm caused by muntjac, but Rob tells me it is important that people realise the initial devastation at Bradfield had as much to do with our own roe deer as any other species. Further confusing things, it is probably true that the roe in Bradfield Woods are not really ours at all.

In southern England, by the late medieval period, roe deer had almost been extirpated due to hunting and deforestation. In the sixteenth century, the few remnant populations that remained were in northern counties running along the border with Scotland. By the 1800s, numbers began to rise but the roe in southern England are all said to be descended from reintroductions: the most notable being 12 deer that were shipped from Württemberg, in the wilds of southern Germany in 1848, to Norfolk's Thetford Forest where Rob lives today.

Like big Jim, Ross Guyton believes that those who stalk are often paradoxically responsible for booming deer numbers. Twenty years ago, he remembers lots of people who were happy to be out for hours on end without seeing a beast, but lately a generation of 'trigger twitchers' has emerged who want to be able to shoot something and have it hanging up in the larder within an hour. Consequently, deer aren't managed effectively, in a way that benefits woodland, but are often only shot when cash changes hands. 'It's perverse,' he tells me bluntly. 'What they're after is a "shooting-fish-in-a-barrel" type of situation when what we need is for stalkers to be thinning the deer right out.'

Until the rise of the biomass boiler, in the mid-2000s, which turns timber into heat, and in doing so, gives woodland some value, Ross had to contend with forestry being thought of as a waste of good ground that would be better off farmed or built on. Back then sporting agents, out to sell hunting, had the ear of landowners. It was a simple case, he recalls, of deer trumping trees when it came to filling estate coffers with cash. The picture is now a brighter one, but Ross feels most people still don't recognise the damage deer cause. 'So often,' he tells me, 'farmers and landowners think they don't have that many deer, so what I do is I crouch down and tell them "I can see 100 metres through this woodland." At which point they crouch down next to me and they go "Bloody hell, you're right." It shouldn't be like that, you shouldn't be able to see any further than 15 metres into a proper wood.'

As well as losing semi-natural woodland habitat, nightingales have lost hundreds of thousands of acres of managed coppice. The ancient coppicing process, a craft that once sustained whole communities, relies on the regenerative ability of broadleaf trees to reach out in search of new life. An area of coppice is separated into coups with the saplings that sprout from cut trunks, known as stools, being harvested on rotation, providing a mosaic of perfect nightingale habitat. It is difficult to estimate just how much coppice has been lost but it is believed that the Romans managed vast swathes of lowland England to provide charcoal for military ironworks. Right up until the 1800s, coppicing remained ubiquitous before beginning a steady decline, and by 1900 this had sped up rapidly due to a collapse in traditional markets for the harvested wood, such as thatching and tanneries which burned faggots to generate smoke. Between 1905 and 1982, the total area of England being actively coppiced fell from almost 600,000 acres to just under 100,000, representing a catastrophic loss of habitat for the nightingale.

The trouble now, Ross explains, is that where any coppice remains, regeneration is often impossible because as soon as the faggots are harvested, deer eat away at the regrowth, at which point the tree attempts to regenerate again, only for the saplings to suffer the same fate. There are only so many times broadleaves can reach out in search of life before they simply give up and eventually, depleted of nutrients, wither and die.

The fire at our feet hisses in the rain and in the high-summer damp smoke hangs heavy. Behind us, bundles of hazel faggots are lashed together, piled up around the entrance to a makeshift shelter where an old radio plays beneath a tarpaulin. Andy Birnie can't remember when he decided he would make his living from the trees but he thinks it probably all began when he came upon a coppicer while running through the woods as a boy. For a while, in the 1990s, he drifted away into community work and taught people who had just come out of prison about birds and conservation, but the woods drew him back in.

'What I like about it,' he tells me, while pushing a piece of unburned hazel into the flames with his boot, 'is that we're maintaining a habitat right here but also that the products I make go into maintaining other habitats.' Every month a local man who restores riverbanks purchases 350 of Andy's hazel faggots, which are used to reconstruct meanders and create silt traps. The technique is an ancient one, and every so often, when rivers run dry, fossilised faggots, dating back to the Roman period, rise up out of the mud. The estate we're on, a couple of miles east of Stockbridge in the Test Valley, retains 200 acres of coppice, which is actively managed by ten men. Since the medieval period, coppicers have paid the landlord by buying the rights to work on 1-acre plots. Currently, Andy only pays £200 an acre, but back when there was more of a market for their products, coppicers would be invited to a local pub where they would bid for the woodland with the straightest faggots, and there are stories of drunken coppicers parting with thousands of pounds.

Andy leans on a tall oak chopping block, shifting his weight restlessly, while trying to guess at just how much woodland has been lost over the past couple of hundred years. 'It's either been grubbed out, destroyed by deer, or it's coppice that's long out of rotation.' He's not keen on pheasant shooting, not least because he's vegan, but he tells me that across Hampshire lots of woodland has only remained where it's used to hold gamebirds. He talks at pace but stops every few sentences to ask me what I think about it all. By the shelter, above the sound of the radio, nesting blue tits are twittering. 'In the pipe,' he says brightly, noticing my distraction. 'There was a great tit nest too in my Kelly kettle, hanging from that tree there, but a squirrel got in and ate all the chicks. They might try again. There's every chance.'

It has been some years since there were nightingales on the estate, but more recently, in another wood, just a couple of miles away, Andy was listening to a radio programme about birdsong at dusk one evening when he heard a nightingale sing. 'I was loading up the Land Rover and they played a recording of one and I suddenly realised a real nightingale in the wood was calling back. I turned off the radio and just listened. It was magical.' He gives the fire a few more prods and I think about that lonely bird from a dwindling population trying to communicate with an electronic recording.

Twenty yards to our left, there's a small coup of stools, some centuries old, all cut just after Christmas, the latest faggots to be harvested as part of Andy's vague seven-year rotation. The books say the number of coups should always be the same as the number of years you plan to allow the faggots to grow, meaning you can work cyclically, harvesting one area every year, while working your way back round to the beginning. But in all the years he's been at it, Andy's never found it to be quite that simple. He kneels and starts pulling back the young leaves around the shoots. 'You see how soft and delicate they are and you can see the deer damage.' Between his finger and thumb, he holds a spindly piece of hazel, eaten away at the end. 'We call a coup a burl in Hampshire,' he says, looking up, 'rather than a coup, a very local word.'

I walk to the edge of the coppiced area and look out over a vast field. 'If you stand there for long enough,' Andy tells me, brushing the red mud off his overalls before gesturing to where woodland becomes plough, 'you see squirrels and jays burying hazelnuts. If you left it, it would become hazel. It all wants to be hazel, really.'

Across the fields, almost at the limit of the human eye's range, where a grassy slope kicks up, a dark shape is moving sedately along the horizon. 'It's a fallow, I think. Wouldn't you say that's a fallow?' I respond to Andy's question with vague noises and we watch the creature wander away over the hill. Andy believes deer are a huge problem. The estate gets paying stalkers in, but none of us coppicers think they hit the deer as hard as they should.

Turning back from the field, I push my way into dense hazel that'll be cut in December when the last leaves are falling. Over the past decade, Andy has noticed how much bolder the muntjac have become. Back in early summer he was sitting in his Land Rover having a cup of tea with one of his customers when one crossed the path right in front of them and the customer wouldn't have it that it wasn't 'a cat with a funny walk'. Andy's stories are told with a blend of wide-eyed zest and nervous uncertainty as though he's never heard them before. Weaving among the coups, he mutters angrily while pulling at the old man's beard, which strangles some of the hazel. He follows the suffocating clematis through the trees, tearing as he goes. When we step into the open, the rain has almost stopped and Andy tells me that beyond the declining demand for faggots, one of the biggest problems facing coppicing is that the community is unwelcoming. 'The trouble,' he explains, as we walk along a track, 'is that a lot of the people who do it always moan that nobody wants to come into it, but when someone new comes along, all they get is criticism and they get watched like a hawk. If they do anything wrong, it's the end of the world.'

The sun comes out, light streams through the oak canopy, and all around golden stripes dance on the hazel soil as the breeze blows among the treetops. Andy and I carry on down what is said to have once been a drover's road before turning into some coups that were last harvested three years ago. Most of the cutting used to get done in the dead of winter when the sap was low, meaning there was time in the summer months for Andy to walk the woods looking at butterflies and peregrines, but the old rhythms of the coppicer's year have become warped. It's so mild now that the sap doesn't really get low, and given the amount of rainfall in winter, Andy's clients who repair rivers can't get onto the fields without getting their machinery stuck. It used to be said that you should stop cutting by Lady Day, 25 March, but the weather is rewriting woodland lore and Andy has to meet the demands of his order book.

The coups around us come up to our waists and Andy tells me that in northeast Scotland they often only ever got this high because they wanted short faggots for weaving lobster pots. 'I like that,' he says, 'the way that the uses for the wood actually shaped the land-scape.' Just ahead of us, a coup rustles, as though it's been hit by a sudden gust of wind and a fallow doe bursts out and bounds away through the hazel. Watching it out of sight, with a smile on his bearded face, Andy explains that often, at this time of year, 'they have a youngster – they'll give birth, hide the fawn, then eat all the coppice round the hiding place in a big circle.' He walks on and I linger a moment to push my hand against the warm earth where the creature lay.

Back by the shelter, the hazel has burned right down but thin smoke is still rising. Andy thinks he'll be coppicing for a while yet. 'I don't really have a pension and when it's good I can sell about 300 pounds' worth of faggots a day, but then of course you sometimes go for weeks without making anything at all.' He splits another piece of wood and places it in the embers. Back when Andy first started, there was an old boy in the woods called Georgie May. Andy kneels down and speaks in between blowing the fire aflame. 'Georgie was a Plymouth Brethren from King's Somborne. "There were 33 hurdle makers in my village, 33 hurdle makers in King's Somborne", he used to tell me.' The hazel takes and Andy stands. 'At 93 years old, he would still come over and he'd say, "This is how you split a rod, boy, I'll show you."' Andy picks his billhook up off an oak stump. 'Georgie used to have his billhooks so long and he sharpened them so much that they looked like sickles. I can't remember when he retired, but right up until he died the gamekeeper here would find him wandering up the track. He'd always say he was just having a look but I guess he missed it. He used to pinch things too, the odd bit of wood.' Andy worries it sounds a bit pretentious but thinks it does get to be like that. 'You spend so long in the same trees that it becomes part of you in some way, or maybe you become part of it.'

When Rob Duncan was a little boy his neighbour, Andrew, told him he would never be good enough. At weekends, when Andrew went off with his father to ring nightingales, Rob stayed in and drew pictures of them. 'He was just really competitive,' Rob says, running his tongue across his thin grey moustache. 'We were best mates, but it had been his thing first.' Rob pauses and blows on the creamy feathered breast of the blinking bird in his hands then turns to me and smiles. 'I suppose he's in the big time now, is Andy. He's like discovering species over in Manaus and he wrote a book called *Birds of Brazil*, but I rang twenty nightingales last year, more probably than anybody else in the country.'

Crouched by his tent, in the dawn dew, James Booty is resting his notepad on his knee. 'Wing length eighty-six was that?' Rob glances down to where his thumbnail sits on the metal ruler. 'Eighty-six,' he confirms, 'that's big, quite big that.' He lifts the nightingale to his lips and blows on it again, the feathers parting to reveal a thin-skinned purple breast. 'You can see he's male because of that cloacal protuberance, that little bulge. Do you see, it's swollen with sperm?' He blows once more then runs the back of his index finger across the feathers.

Five years ago, at the age of 55, Rob retired as a primary school headmaster. Most of his days are now taken up with looking after his wife, but whenever he can, he's out ringing. As a boy, in Birmingham, Rob used to dream of nightingales, then in his early twenties he moved east to Suffolk in order to be in the heart of nightingale country, and a few years later he finally ringed his first. 'I was 27 and I was trembling. It was more perfect than I'd ever imagined it would be.' With terror in its voice, like a rabbit caught by a hawk, the nightingale, feet held between Rob's fingers, starts squealing. 'It's okay,' he whispers to it, 'you're going to be okay.' When the bird has closed its beak, brown on the outside and orange within, Rob passes it to James who places it in a small paper cup on a set of weighing scales.

Hands resting on his thighs, Rob watches the procedure as he talks. James is still ringing on a British Trust for Ornithology C permit, which means that Rob, as an A permit holder, needs to supervise. 'People sometimes ask why I'm still at it,' Rob tells me, 'but it never stops. No two nightingales are the same.' The rusty brown tail feathers are motionless and at a glance you'd think the bird was dead. '21.24 grams,' James says excitedly, while penciling the number down in his notebook. 'That's a nice healthy weight for this time of year in peak breeding,' Rob replies. 'But in autumn before it heads back to Africa,' he adds, turning to me, 'it could be up to 23 or 24.' James picks the cup up with his right hand and tips the bird out onto his left before reaching for his banding pliers and a tag. 'JX5156' – he reads the number out as he squeezes the small piece of aluminium around the bird's leg. 'What drives us,' Rob tells me, as James passes the bird back to him, 'is it means we get to see where nightingales go. It's astounding. This little bird could be killed by a hawk or shot on migration or it could die in a sandstorm in the desert, but if it isn't and it's caught by a ringer elsewhere, it'll help reveal what they're doing, and more than ever we've got to know.'

The grass is thick and wet and as we walk through the heavy grey air, the damp soaks into my boots. All night it poured and it'll start again soon, but for a few cold hours the rain has stopped. At the edge of the meadow, on all sides, bramble and thorn, flowing together, sprawls 15 feet high. 'We've got to make people think differently,' Rob tells me, as we cut left towards the nightingale's nest. 'We've still got this obsession with slash and burn.' Rob slips into a high sardonic voice. 'It's prickly, it's ugly, it's invasive. We've got to keep England tidy.' He turns and looks at me, his blue eyes wide with boyish exasperation. 'I see this and I just think it's gorgeous. Look at that blackthorn. This was a farm once and when the owner died they left it in trust to run wild for conservation. There are seven pairs of nightingales here, seven pairs.' Two years ago, in an attempt to do his bit for a bird that has given him so much, Rob founded ScrubUpBritain. 'All one word,' he explains. 'No spaces, the name was my idea, and we're not an organisation. We're just a ragtag bunch of guys, professional ornithologists and

24

amateurs, who are trying our best to make people understand what's actually happening to these birds.' James glances at me and smiles. 'Rob's sort of our spokesman.'

In front of us, running between a gap in the thorns, a length of mist netting is held up between two metal posts. Twenty-six minutes previously, the bird in Rob's hand flew into the baggy nylon mesh, whereupon James untangled it. Ringers are only able to keep a bird for half an hour and it is time for the cock nightingale to go back to its nest. 'He'll have been born here,' Rob tells me as he holds the bird up in front of his face. 'Not just here, but after returning from Africa he'd have come back to the same bush he fledged in. They always do.' Rob nods, open-mouthed, when I ask him where in Africa they go. 'That's one of the things we don't really know yet and the more I learn the more I want to know. It could be the Ivory Coast. It could be Ghana. We can't just protect them here on their breeding grounds. We need to know exactly where they're going and where they stop over and that's part of ringing.' Rob blows on the bird's chest one last time. 'Can you see it's got no fat? The males come in first, around mid-April, and then they sing to bring in a mate. I would imagine most of the nightingales around us will have brought in a female by now. At the moment he doesn't want fat because he has to stay nimble while he's helping to raise young.' Rob stands at my shoulder and tells me to cup my hands. 'Just let it do what it wants. Sometimes, they stay for a moment.' With the bird's head held between his fingers, he places it softly down. Its four little toes stand light on my palm and the bird lifts its tail out behind it while pushing its beak into the air. 'People say they look boring,' Rob whispers, 'but those people are crazy. Just look at that orange and rufous and cream.' I can feel Rob's breath on my face and then the bird is gone, fluttering up and out into a blur, across the field in the damp dawn.

At seven, the sky darkens, the temperature drops, and the wind starts to lift. As James takes the net down, Rob and I go in search of a flask of coffee he left under a bush somewhere. 'This is just perfect,' he tells me, as we head back the way we came. 'I wonder whether the ace card of this site is that it's got so much grass meadow and so many flowers that it gives the deer plenty to eat without them having to browse out thorns and bramble.' Rob stops and crouches next to a patch of nettles. 'They love this dense tapered fringe. When females build their nests it can be just in from the edge. You've got all that lush green with caterpillars and larvae. All that food.' Somewhere up ahead of us, cutting Rob off and drowning out the blackbirds, a nightingale begins, sharp jaunty notes at first, running out into bursts of high rolling song. After it ends, Rob turns in a circle, looking round and round, waiting for the bird to sing again. 'That was the sound I dreamed about as a boy,' he says eventually. 'We cannot lose that. We've got to do everything we can. To lose that, for me, would just be the end.'

'How does one actually do this?' Beneath the bough of a fruitless cherry tree in the garden of The Spaniards Inn, Sam Lee, the most indefatigable force in the British folk scene, sits across from me. At 20 he was dancing in Soho and by 30 he was on the road, collecting songs from travellers and gypsies. Lately, he's become a writer, an activist, an award-winning musician, and he cannot open his ketchup. 'I think you just tear the top off,' I reply. He stares intently at the little packet then rips it with his teeth. He arrived late by bike, pedalling hard over Hampstead Heath, and sweat is pooling on his forehead beneath silver-flecked chestnut curls. 'Shall I do it like this?' On a slate in front of him, he squirts the ketchup in a circle and then turns his attention to the mayonnaise. He seems satisfied with the solution and we start to eat our chips.

Every year, in spring, Sam heads down to the woods on the south coast where he spends six weeks living among the nightingales. He listens to them, he leads people out into the darkness to hear their music, and he duets with them. Sam tells me that, creatively, nightingales mean more to him than I could possibly imagine. 'They taught us so much of our culture and they gave us the rhythms of our lives. They held our stories, they held our narratives, and they held our mythologies.'

Sam believes we have come to a dissociative juncture where birdsong is still such an intrinsic part of our inheritance but we've become deaf to the language of the landscape. While he realises his wanderings might seem eccentric, he thinks the numinous emotions that are evoked when he leads people through the trees to hear the nightingale's chorus make perfect sense, 'because birdsong still feels so much part of us and it has so much purpose.' What the listeners he takes with him appear to be experiencing is a sort of completion and fulfilment that gets lost when we're adrift in the riptide of modern digitised life.

After countless hours of listening to nightingales, communing with them through song, and studying their endless guises in music, art and literature, Sam sees them as embodying so much more than merrie olde England. He sips his shandy and speaks carefully: 'I love the way so many writers and artists have approached the bird differently. Keats's 'Ode' and then the kind of wonderful response from DH Lawrence on Keats's woefulness. It gets to it.' The riposte Sam is referring to appeared in the September 1927 issue of the now-defunct *Forum Magazine*, after Lawrence heard nightingales while staying at a villa in Tuscany and felt sure that the young poet had got it all wrong.

How John Keats managed to begin his 'Ode to a Nightingale'
with: 'My heart aches, and a drowsy numbness pains my senses,' is
a mystery to anybody acquainted with the actual song. You hear the
nightin-gale silverily shouting: 'What? What? What, John? Heart
aches and a drowsy numbness pains? tra-la-la! tri-li-lilylilylilylily!'
And why the Greeks said he, or she, was sobbing in a bush for a lost
lover, again I don't know. 'Jug-jug-jug!' say the medieval writers, to
represent the rolling of the little balls of lightning in the
nightingale's throat. A wild, rich sound, richer than the eyes in a
peacock's tail.

Sam thinks that John Clare, the nineteenth-century farm labourer, whose work wasn't much appreciated until well after he died in a lunatic asylum, probably wrote the nightingale's song most accurately, albeit with occasional 'dispassion'. In Clare's *The Nightingale's Nest*, a man takes a companion 'up a green woodland ride' to hear the 'rich ecstasy' of the bird pouring out its heart. When they find the nest, hidden among the bramble, it is as though a spell is broken and the bird stops singing. For all Sam believes that Clare paints it right, he doesn't believe Keats or Lawrence got it wrong because the nightingale 'is a wonderful canvas and a mirror for the listener', a muse of endless and kaleidoscopic inspiration.

Two whippets, one young and blue-eyed, the other old and grey – both wearing turtleneck jumpers – wander past and sniff at Sam's ankles. 'There's a real type of clientele here,' he whispers with a smile, before his mouth becomes serious again. 'Some nights I go and the bird is ecstasy and exuberance and sometimes it's heartbreaking.' Sam believes that without birds, there would be no music in the way we hear it. 'There's never been art without birdsong. The first evidence of music being played was on bird bone flutes. Ask me, would I rather never hear a nightingale again or listen to

Joni Mitchell again, that would be hard to answer because they're all dependent on each other.' He dips a chip in the mayonnaise, bites the end off, and looks up at me as though trying to sense if I believe it all.

A wasp, having been batted away from every table in the garden, drifts down over my beer and settles on the rim of the glass. Sam is interested in the way nightingales correspond with each other through song and he believes on an emotional plane that this correspondence extends to humans. 'It's a species that divines out of you, drawing up emotions. It almost works on a medicinal level.' For all Sam's flirtatious mystique, he tells me, between mouthfuls, that we need to be doing everything we can to make the battle against extinction inclusive. 'We're reaching a stage, certainly with ecology, where the most dangerous thing is to turn people away from the issue.' He moves forward in his chair before continuing. 'People simply start to say I don't like all this stuff, all this Extinction Rebellion stuff, sometimes the very people we need.' Sam thinks the power of music can be harnessed to bring the outliers in and to make people receptive. 'That's why it's been at the heart of all great revolutions. It's a dangerous and powerful weapon. You don't change people's minds through information. You change it through the heart.'

Sam is hitching a lift to the Ashdown Forest early the following morning and he needs to go and pack. We wander out through the garden and I look round, wondering where it was that Keats is said to have written his immortal 'Ode' after hearing the voice of that now-vanishing bird. At the front of the pub, Sam unlocks his bike, bumps his elbow against mine, then pedals off up the hill, weaving in and out of the cars before disappearing among the buses.

Beauty beneath his boot

Until we lived the water, river, beck and sike
pond, estuary, bog-peaty-pools
rightly she ranked us gowks, fools
overlooking sense and truth until we stumbled
on the spell, stilled every reed in the marsh.

Colin Simms, 'A First Harrier',
Hen Harrier Poems, 2015

An echo rolled out across the charred mosaic hills as a buzzard fell dying through the branches. I was the child in the passenger seat and the gamekeeper sitting next to me – his smoking gun resting on the wing mirror – wasn't much older, just a boy doing right by the boss. We were on the rounds. Through the woods with a lamp lighting up the treetops, eyes wide for any raptors that had drifted down off the moors to roost after a hard day's hunting. Not quite gone midnight, we had three in the back of the truck and then we drove out into the open where he brought his boot down on the still-warm bodies, stamping them into a bog.

Between 1837 and 1840 on the Glengarry Estate in Inverness-shire, it's claimed that gamekeepers killed 285 common buzzards,

63 goshawks, 98 peregrines, 27 white-tailed sea eagles, 15 golden eagles, 18 ospreys, 6 gyrfalcons, 11 hobbies, 275 kites, 371 rough-legged buzzards, 462 kestrels, 78 merlins, 63 hen harriers and 7 orange-legged falcons. The tally is often bandied about and is always cited as being deeply shameful, but in the mid-nineteenth century it represented a triumph. The numbers are attributed to the Glengarry Estate's manager who had given the list to the seemingly impressed editor of the *Inverness Courier*. Retrospectively, the number of rough-legged buzzards killed is remarkable as only a few dozen winter in Britain each year. Equally, quite how they killed seven orange-legged falcons when they are a rare vagrant, which only occasionally drifts across from Eastern Europe, will never be known. Admittedly, in the early nineteenth century bird identification was less refined and it's possible that all those rough-legged buzzards in the stink pit were actually something else. Equally, though, the tally is possibly best read as a fanciful statement of grisly aspiration. While the details might be complex, the principle was simple – anything that ate game was a problem and all problems had to be eradicated. The landmark 1954 Protection of Birds Act made it illegal to kill all raptors with the exception of the sparrowhawk, which didn't gain protected status until eight years later, but as my memory attests, protection is often just a nice idea.

Of that list, no predator evokes more ire than the hen harrier. A week after that night-time spree, I met a lad who'd lost his job on an estate further north. 'We sit out for hen harriers here,' the head-keeper had said, just a couple of days after taking him on, at which point the boy fetched his dog and caught the bus home.

According to most estimates, there are now fewer than 700 breeding pairs of hen harriers left in the country, representing a decline of almost 30 per cent since the early 2000s. Habitat loss and nest predation by foxes are significant factors, but crucially, it has been found that tagged hen harriers are 10 times more likely to disappear on land managed for driven grouse shooting. The impact these hard-hunting birds have is heavy, with studies showing that moors with breeding harriers produce roughly 17 per cent fewer young grouse, and from weeks two to eight of chicks being born, harrier predation accounts for over 90 per cent of observed losses. In other words, if you're gunning to keep a commercial grouse-shooting venture afloat, every hen harrier is a sharp-beaked business risk.

During the softest March I remember, just after my mother came home from hospital and her brother was found dead after an overdose (I never managed to return his illustrated guide to the birds of Japan), I went to spend a week in a cave on an island in the Firth of Lorne.

I caught trout in the hills, pulled up an empty lobster pot each morning and picked my way among abandoned crofts, old gravestones and a tumbledown chapel on the eastern shore. Modern life grinds on at such a pace that you run every day, forever keeping up, and never stop to consider anything. On an island, awake when the sun comes up, the days seem to last forever and you stand to face it all. The trout in the lochens must have been stocked at some point, probably in the Victorian period, as they were far bigger and more numerous than you would find naturally in a place so devoid of food. One evening, after a successful couple of hours casting a line, I was sitting down on the beach, frying some fish in butter. When it's just you and your dog, you start to rely on their senses. When they run out of your cave barking in the night you grow fearful, and when they sniff the breeze and growl, your hackles stand up too. As the trout's skin turned golden and the unpicked bones started to prick up out of the flesh, I noticed that my dog was looking up between the peaks behind me. I could see nothing so I returned to my fish, but her gaze stayed fixed on the heathery horizon. My stomach tightened and I followed her line of sight before picking up my binoculars. Quartering low over the grass, tacking to and fro on the wind, was a female hen harrier, mottled ringtail with accusatory eyes, hunting voles in the evening sun. Perched on an upturned sun-bleached fish box, I watched her float into view, skimming circles through the air. When she drew nearer, I placed the binoculars down on the shingle. Forty yards away, she drifted by before curling away round the headland, casting a shadow over the sea.

I've never slept as badly as I did on Scarba. I had mistakenly brought a sleeping bag with me that belonged to my brother when he was young and it came up just below my nipples. Every night at around eight o'clock when it started to get dark, I turned in and lay there, listening to the waves breaking over the rocks. Eventually, I would fall asleep until some sound woke me and I'd find, on feeling among the dried goat shit for my watch, that just half an hour had passed. Eventually, in the distance, the sun would appear on the horizon and I'd struggle free from the child's sleeping bag and head out from the cave to collect dead heather for a fire.

That night, as I lay there, I thought about the ringtail sweeping down on the breeze and my mind turned to the boot coming down on those birds, bloody feather and broken bone stamped beneath the moss.

In a lead-mining village, on the edge of the Pennine Way, lives an irascible old poet. Colin Simms was born in 1939 and is seen as the heir to his mentor, Basil Bunting. Bunting was a spy, a balloon operator, a music critic, a conscientious objector and one of the most important poets in the modernist tradition. A vivid and disquieting reader of his work, Bunting wielded the power of sound to extraordinary effect, which echoes powerfully in Simms's own poetry.

Hugh MacDiarmid, the esoteric force behind Scottish modernism, believed Simms to be 'one of the few poets writing proper scientific stuff'. In 2015, 270 of Simms's hen harrier poems were published in a single volume – it is an intimate collection born out of a life spent hunkered down behind dykes and in ditches, waiting and watching and listening. Simms, his publisher told me when I wrote to ask how I could track him down, is exceptionally hard to get hold of – 'No phone, no email, no computer, nada. You can write to him via snail-mail. I suspect he'll be okay with what you're doing but he can be prickly, especially with some of the official conservation people.'

The road goes right, rolling down to Penrith, but I turn sharply left and wind my way up over the top of the pass – Alston, 5 miles. Garrigill is shut up when I get there. No lights on in the village shop, and the George and Dragon is closed. After walking up and down the main street three or four times, a smart lady, dead-heading roses with a pair of secateurs, asks sharply if I'm after something. 'Simms, I'm looking for Simms, Crossfell Cottage.' She looks confused and tells me she doesn't know all the holiday cottages. 'It's not a holiday,' I reply, hearing myself nervously making little sense. 'It's about harriers – his publisher gave me the address.' I say the name again, 'Colin Simms'. She looks at once startled and pleased. 'Colin? Colin's up there, turn right. It's the one on the left. Covered in all the foliage.'

It's a lane I've already been up twice, but the conversation was strange and I want to get away. At the top of the street where it forks, a ginger squirrel dashes across the stony earth in front of me and then disappears into a mass of scrub next to a small austere chapel. When I get to where I lost sight of it, I notice there is an opening and I push through to find it becomes a path. On all sides and above, sprawling shrubs block out the light of the hot July sun and through a window, on a ledge covered in flies, a glass holds paint brushes. I knock and wait and knock again.

Maybe I would rather he wasn't in, but I've come a long way so I linger, tapping my foot. Then I knock harder and press my ear to the door, listening for footsteps. My chest hangs heavy and when I step back, I notice a note wedged beneath the handle: 'Colin doesn't need anything'. I turn and go, feeling everyone in their houses is watching me as I retreat down the street. Maybe you shouldn't go looking for poets who don't want to be found. Back in my car, I look up at the hills on both sides and I realise it's not true. Colin needs all of this, this landscape and the hen harriers still living here. I scribble a note: 'Colin, I wrote some months ago and I came today but you were gone. There was a red squirrel in your garden. I wanted to know more about hen harriers and the time you went for a walk with Hugh MacDiarmid, when he pissed on the dyke where the council built a bench.' Jumping out of the car, I push the scrap of paper under the doormat of the lady with the secateurs, hoping she might put in some sort of good word, and then I drive north.

Cumbria is still full of rabbits and I swerve all over the road counting them. Everyone I pass is Colin Simms or might be: a man on a bicycle and a topless walker in a woollen hat. There are two tiny rabbits grazing beneath a cattle trough, and I drift over to the wrong side of the road to get a better look. A bearded man in a white van coming the other way beeps his horn and curls his hand, pumping it wildly in the air. His thin lips articulate the word, a voluptuous W and a concrete K. Maybe not Colin.

On Hartside summit, the ice cream van has gone but I sit in my car looking down towards where moorland becomes field. Below me, 10 grouse butts are gouged out of the hillside. I reach for Colin's book and scan the index. 'A First Harrier' – it is the poem I wanted to ask about. In his own conception, there is a fastidious 'isness' to his writing. Colin is an observer rather than a projector. The rhythms of his verse reflect the flight patterns of hen harriers. They are birds, not – as is the way with so much contemporary nature writing – canvasses for every human sentiment. But 'A First Harrier' is different. In it, Colin recalls his aunts, when he was a boy, thinking he was a fool who overlooked 'sense and truth' until he stumbled on the 'spell' – seeing his first ringtail was transformative and gave him a different type of knowledge. It was an encounter that granted him a new way of seeing.

A grey fug hangs heavy over Ilkley Moor, torn through by an electric pink gash. It's impossible to say where the peaks on the other side of the valley end and the sky begins. They said on the radio, there'll be thunder in the morning. 'Buddy . . .' The word comes as an intrusion and I flinch before collecting myself. Turning, I find there is a crotch in my face and I look up, taking in a ripe belly. 'Buddy, I'm on a date – I couldn't book a table.' He glances at my plate. A lonely onion ring is drowning in tartare sauce. 'It's a first date.' I hold my hand out to catch the waiter's attention, swept up in the urgency of the moment. 'Can I have the bill?' I mouth, waving my hand in the air with an imaginary pen. 'Thank you, buddy.' He's younger than me and he sounds like he might cry. I pay and walk away over the grass, counting a hundred steps before looking back.

The story goes that a long time ago, when there were still plenty of dragons, Rombald the Giant was running away from his wife across the moor. She had gathered up stones in her skirt and was throwing them at his head. As he crashed through the heather, his heavy foot came down on a great boulder which smashed in two. The larger part came to be known as 'the cow' and the smaller part, 'the calf'. People in twos and threes are spread out across the cow, watching the pink sky soften. On my left, girls in running gear are sharing a bottle of rosé, and above me, three teenage boys are passing a spliff back and forth. It is the eve of the Glorious Twelfth, 9 p.m., still 28 degrees. Beneath the rocks a lost lamb, almost old enough for slaughter, is bleating and further down, on the road rising up out of Ilkley, a man in lycra is fighting a bicycle, weaving all over the road, churning a big gear.

There are at least 400 Neolithic 'cup and ring' markings on the rocks across Ilkley Moor. Sometimes a straight groove runs through the centre and sometimes, like a cut tree trunk, the rings spread concentrically outwards. Nobody knows who carved them or why, but many believe they have some sort of mystical significance and are the work of people who knew there to be something greater.

The pink streaks across the sky are fading to that pale bluish grey of the male hen harrier's wings. Above me, the boys are getting high. 'They get so fucked up in the wash,' one of them says. 'I told my mum not to put them in the washing machine.' I look up to see that the other two boys aren't listening. The one nearest me keeps saying he's a Viking and the third is wide-eyed, looking out into the distance. 'Shut up and tell me that this is not God's own land,' he says suddenly. They stop talking and turn to look at him – he is smiling, transfixed. They don't respond. All around me, in the half-light, there are faces gazing upwards with the same devotion.

For a moment, I think I hear a cock grouse cackling on the wind, but three mopeds whine their way up the road, drowning out any sound. Beside me, carved into the rockface, are two names: 'J Whitehouse, May 1875' and 'E M Lancaster, 1881'. I watch the red tail-lights of the bikes disappearing over the hill on the moor road and I wonder whether, in 12,000 years' time, people will look back on those names as we look back on the 'cup and ring' markings, or whether there will be nobody left to look back at all. It seems we've always needed it. In the Neolithic period they carved circles in the rocks and faith was available to Whitehouse and Lancaster, in Christian abundance, but just a couple of decades later, the tide went out. I look from the sky to the people's faces and I wonder, in a secular, digital age, if we are now turning to the wilds for faith and silence, but the gods in the hills and hedgerows, the hen harriers and the turtledoves are fading too.

Mass is ended and the congregation drifts down off the rocks. 'What you got a notepad for?' a boy in his late teens is shouting across at me. 'Are you a copper?' On my left the hill drops away steeply. 'It's for a book about birds and art and how birds make us human and stuff,' I reply. The boy shouts it – 'Big lad's writing about birds!' – the people he's with start laughing. He turns back and lowers his voice. 'Nobody wants to read about birds. People want to read about drugs and sex and stuff.' He draws hard on his vape and exhales. 'Write about that, lad, and put me in it.'

An old shooting hut rusts away in a stony scrape on the most southerly edge of Ilkley Moor. Across the country, the first grouse of the season are being driven to waiting guns. 'Do you want half a scone? It's fruit.' I hold out a brown paper bag and the blond-haired boy in black brogues eyes it strangely. 'I'm cool, thanks.' As a child in the 1990s, Luke Steele spent his weekends being taken along to demonstrations by his parents and he discovered the fight for animal rights at thirteen. I pick the raisins out and he speaks heavily, as though a crowd is there to listen. 'This is where the grouse moors start. The firewall as it's known, from Otley up to Greenhow Hill. The most easterly grouse moors in the Pennines. When raptors hit here, they get shot and poisoned.'

Out in front of us, a red kite, swept on an uprush in the close morning air, catches Luke's gaze. As he watches it, he tells me that over the past few years he's devoted himself to trying to uncover the truth behind over 200 cases of disappearing birds of prey. 'We might see a hen harrier today but it's unlikely.' He smiles wryly as he says it and fixes me with small black pupils. As far as Luke knows, there were only two hen harrier nests on Nidderdale over the summer. It could mean 10 chicks and he's been told a clutch has just hatched, but the survival of the young birds will depend on their parents' ability to feed them. 'A lot depends on the vole year,' he continues. 'It's been a good vole year.'

Luke turns away and goes back to watching the kite. He explains, as I pour myself a cup of oily tea from my flask, that 'there's no official estimate but there are around forty moors in Nidderdale and each one is capable of having a breeding pair.' The numbers start running through my head. 'So that means there could be well over a hundred chicks a year?' 'Forty breeding pairs,' he replies with a shrug. 'It's got that capacity but obviously it's also a bit dependent on how much food there is for them.' Luke tells me that in the 25 years he's walked the hills of Nidderdale, hen harrier persecution has intensified as people have pushed to shoot more and more grouse. 'You only have to go up to the Orkney Islands, where you see hen harriers exceptionally frequently – there's no grouse moors and they're thriving.'

I screw the cup back on top of my flask and Luke points to a copse in the distance where a badger died after being caught in an illegally set snare. Luke's lips are dry and cracked and he chews at them as he speaks. 'They know they shouldn't be doing it. Just like they know they shouldn't be persecuting hen harriers, but there's an expectation. Whether it's spoken or just cultural pressure, most gamekeepers know if they don't produce the grouse, they're out.' He pauses as though ordering his thoughts and then asks me how big a problem I think it is.

It is not yet ten but the air is hot and dry, and as we wander back down a dusty track towards the car, sweat runs into my shoes. In the distance, below us to the right, there is a rectangular block of trees. 'Do you see that?' I ask Luke. 'Do you think that if grouse shooting was to go completely, this whole landscape might be turned over to commercial timber production and those two pairs of hen harriers hanging on, as well as lapwings and ouzels, would lose their habitat?' It's a question he's seemingly answered before. 'You can't just leave moorland unmanaged, but you need to look at the bigger picture of how you bring money in.' I tell him all sorts of people have told me that income from tourism could never really cover what would be lost if grouse shooting was to go. He moves his head from side to side, neither in agreement nor disagreement, before telling me that large swathes of grouse moor are blanket bog, which provides an important carbon sink, but just 4 per cent of it in Britain is deemed to be in a satisfactory state. 'There are a lot of businesses around Yorkshire that would quite happily invest money into restoring peatland to offset their carbon footprint.'

Out to our right, 20 yards beyond the path, three grouse burst from the blooming heather and disappear over the horizon. I watch them, thinking about all the times I've seen that sight followed by the sound of guns and a bird tumbling to the ground. Luke has wandered on ahead and I shout after him, asking whether he'd be okay with people hunting a few grouse over spaniels for their dinner.

'It would be acceptable in terms of progress,' he replies. 'But when you say progress,' I shout back, over the sound of a lorry grinding up the road, 'do you mean that you would like to be in a situation in which people are only allowed to shoot small numbers of walked-up grouse, because from there you might be able to get to a point where nobody is allowed to shoot grouse at all?' He tells me he sees what I'm getting at, 'but we're in this place now where we've got an absence of hen harriers and it's driven by the kind of dominant desire for big bags and lots of grouse'.

On 3 October 1957, four men serving as beaters on Pockstone Moor were blown up, mid-drive, when a World War II shell exploded. We sit on a grassy knoll, looking out over the sweeping silence. Luke explains that the roosts are on the north side of Nidderdale but the hen harriers fly down over Pockstone and the flank in front of us is where several tagged birds have gone missing.

I pull my glasses down onto the tip of my nose and look through my binoculars towards where the ground rises up, not sure exactly what I'm hoping to see. 'How many birds?' I ask. 'Five,' he replies, 'at least five in this area of the valley.' A long way out, a magpie on a stone dyke is picking at something dry and dead. 'Any prosecutions?' Luke shakes his head. 'Look at this valley, you've got hen harriers that have disappeared here. You've got one over there. Behind us you've got an estate where two dogs were poisoned after eating bait. You've got disappearing hen harriers on the other side. Then down in Pateley Bridge, you've got buzzards dropping dead in people's gardens.' I tell him I really feel that everybody I know who shoots is opposed to hen harrier persecution. The polished words tumble out of my mouth – I've said them so many times it feels as though they aren't mine anymore.

Luke smiles. 'This is one of the things. People like yourself are saying this is wrong but it's kind of missing half the argument, which is that there is a solution to this. Licensing grouse moors wouldn't be a panacea but it would be proper progress.' The rasping guttural call of an old cock bird can be heard in the thick warm air. Luke pauses for a moment as though to hear it out. 'We need to be able to say if you want to shoot grouse, you have to show you're preserving birds of prey. Licensing should be viewed as positive by all, as it takes the incentive for killing hen harriers and turns it into an incentive for saving them.'

Luke accepts that hen harriers reduce grouse stocks to the extent that large commercial operators might find their ventures are no longer viable, but he believes there are simple choices to be made. 'If birds of prey continue being persecuted, there will be no more grouse shooting of any sort. The voices are getting louder and louder.' I look around me at the tussocky cairns pockmarking the landscape and wonder who built them and why. As part of a small team, who work largely under his direction, Luke sets camera traps across the moors in the hope of catching keepers in the act. I tell him in some ways it all seems a bit easy. 'It's like busting a runner with 3 grams of coke in his sock on a Saturday night. It makes no odds to the kingpin.' He holds out his hands and nods. 'You're right. We haven't seen the police going for the top guys in terms of grouse shooting in the way they go for dealers, but the structure is very similar. You've got some young lad fresh out of college, given the promise of a really good life. He's compelled to do this stuff and it'd be him who pays the price.' Thirty yards to Luke's right, a young rabbit has emerged from a hole beneath the remains of a wall. It eats contentedly, button-black eyes glinting in the sun.

'Can I get scraps?' Luke is speaking in his chip-shop voice. 'Chuck a load on.' The man in the white coat behind the counter scrapes his scoop across the bottom of the cabinet, gathering fragments of frazzled batter, then drops them in a heap on top of the mushy peas. By the time we've walked down Pateley Bridge High Street and are sitting on a wall, overlooking the Nidd, my Coke is already warm.

I open the polystyrene box, break the cod in two, and start to eat. 'So, when you get tip-offs about hen harriers being killed, where does that information come from?' Luke looks up from his lunch, a large chip skewered on a wooden fork between his fingers. 'All sorts of places, sometimes walkers, sometimes runners but gamekeepers too. It's a small section but it's very significant.' I am trying to dislodge a fish bone from between my teeth with my tongue and it muffles the dubiousness of my reply. 'Gamekeepers, what sort?' Luke laughs. 'Well, they're local grouse keepers obviously, usually older ones, but I can't really say. There aren't many people they can tell without putting their jobs at risk. Sometimes they notice something or they hear something down here, something in the pub.' He looks up towards the town's steep streets.

Out over the hills, the first clap of thunder rolls. 'When I said to someone, a keeper in Leicestershire, that I was coming to see you today, he told me to keep one hand over my tea.' Luke tries a smile. 'I wouldn't have ever done that.' He eats fast. Most of his chips are gone. 'I found a list,' I continue, 'from when you were in prison, put together by an animal rights group. It had an address. People could send you things, writing materials, biscuits, stamps.'

There are children playing in the river and a little girl is throwing stones for a spaniel. Every time one splashes into the water, the dog dives down and emerges with nothing. 'Yeah, that's fine,' Luke says with a nod. I look down at my fingers. They shine with grease and ketchup. 'Things I've read seem to vary on how many prison stints you've done. Some say two, others three?' Luke folds up his poly-styrene box slowly and places it on the wall before answering. 'One was for setting up a protest camp, the other was at a laboratory breeding farm. It wasn't violent.'

I want to ask about reports that a group he led gathered the details of scientists they were harassing in order to cook up and spread false stories about them being paedophiles, but hooves are clipping somewhere behind us and Luke turns towards the sound. He likes horses. The noise grows louder and louder and then they appear, sweat beneath their girths lathering up into a foam and their mouths open, breathing hard in the heat. The riders kick on, trotting up the bridge over the Nidd and away, the sound of the hooves on tarmac fading beneath the gush of water.

On our way out of town, heading up into the hills, we pass a broken-down car towing a caravan. Its owners, t-shirts sodden with sweat, stand in the shade of a roadside rowan, passing a 2-litre bottle of Pepsi back and forth.

A welcome breeze blows the sweet tang of bracken in my face and Luke is breathing hard behind me. We are clambering up the north side of Ilkley Moor towards Woodhouse Crag. When we hit the path, we stop and look back down into the valley. The sky is growing darker and the houses are shrouded in a charged fog. Luke turns and walks ahead of me, casting his eyes across the hill. Above us, there is an outline of a track running through the heather and I climb onto a wall to get a better look. 'Do you think it could be up there?' He glances over but tells me it's more obvious than that. Up ahead, on the path, a topless man, long hair and khaki shorts, is walking towards us, a plastic bag on the crook of his arm. I ask him if we're on the right path to the stone. 'Up there, just on the other side, over the fence. Three minutes, over the fence. It's a copy. It's the other one you want. The proper one. It's behind.' He seems strung out, and his eyes, as he talks, make it look as though he could go on until tomorrow. We thank him and keep walking. Luke points to the horizon. 'That's the place I mentioned, the ridge where the bird disappeared.' He starts telling me that we should see hen harriers as being like canaries in a coal mine. When our ecosystems are rotten, they come up dead. I am half listening, half looking at the raven tattooed on his arm. Crisp black wings, it looks new, and below on his wrist there is faded ink, writing mostly gone. 'A few seasons ago,' he continues, 'the shooting here stopped and the short-eared owl came back and there were hen harriers here in winter too.'

Beyond the next wall, up over a style, dark painted railings come into view and Luke heads towards them. For a moment we stand in silence, next to each other, peering down at the four spiral arms with a cup in the loop of each, carved in the rockface. For as long as anyone can remember, it has been called the swastika stone. Like every other carving on the moor, what it means is unknown but it may be significant that across various ancient civilisations the swastika was symbolic of the sun. The railings are as tall as I am and I push my face up against them but I can't see another. 'Is that the copy?' Luke shakes his head. 'I'm not sure.' It's too tall to climb, so out to the side, then backwards over a boulder, I start to descend, feeling beneath me with my toe for a solid hold. Ten feet down, I traverse round the crags until I can reach up and pull myself onto the slab. On all fours, I run my fingers over the granite, trying to feel the markings made by Neolithic hands. Luke stands above me, looking down, his fingers clasped round the railings. I learn later, from pictures, that the carving was just beneath my feet, but it's grown so faint now that you can hardly tell it's there.

Tumbling down over red rocks above us, the burn burbles a song of half-heard words. Thin grey moustache and green shorts pulled up over his belly, Martin Davison steps out from beneath a twisted downy birch and plunges into the water. It must have rained in the night. The water rises, he tells me, along the English border high in the hills and eventually merges into the Tyne. 'It's really about three or four different burns,' he says, when he gets to the other side. 'If you like, I can tell you all the names but you can't let on to anyone because nobody is meant to know where these birds are.' Cold clear water runs around my calves and my cock loses an inch. 'You'll melt in that jumper,' Martin says as I wade towards him. 'It's the best summer we've had for donkey's years. But mind you, we had one of the worst springs.' When I've made it across, I take the jumper off and tie it around my waist. Behind me, Stephen Murphy has found a drier route round and behind him, camouflage bucket hat on his head and a large, faded rose tattooed across his right hand, Gavin Craggs looks past us all, up to the rocks beyond.

Stephen pushes on ahead, picking his way through the knee-high heather. A small wiry man, now 58 and Natural England's leading advisor on hen harriers, he was working in a photography shop in Liverpool until he was 35. 'It was like a photographic development place,' he tells me with a shrug, when I catch up. 'I'd done it since I was a kid and one day I just came to the conclusion that all I was doing was making the bossman a bit of money every year, so I just came out of it and did a degree.' After his Bachelor's, Stephen went on to do a Master's and then progressed to a PhD on the findings of Natural England's first hen harrier tracking programme. Nobody seems to be very sure why it was never finished. Some suggest that in one way or another the findings were going to be excoriating and pressure was applied, but all he tells me is that he was hit with a barrage of Freedom of Information requests, at one point 37 in a day, and it made things difficult.

Stopping a moment by a broken sheep shelter, Stephen calls down the hill, asking if we're still on the right track. 'It's just to the left of that patch of bracken this year,' Martin replies. 'We had some forestry work down the bottom there and I think it's pushed them along ever so slightly.' Martin grew up in the hills around Kielder and has stayed among them all his working life. 'My study area is 50 square miles,' he tells me when he gets to us. 'Basically, I just locate as many birds of prey as I can, then I tell Forestry England how to avoid conflict with them.' He pauses to watch a caterpillar crawl up the stem of a fern in front of us before adding, 'owls obviously too'.

For 50 yards or so, we trudge on and then Martin touches my shoulder. 'That's what we want to hear,' he whispers, pointing to the horizon. 'That's the noise that tells us everything's okay.' Two hundred yards ahead, up above the heather and beneath a thin wisp of cloud, a hen harrier, chittering sweetly to her chicks, cuts circles through the air. 'You hear it, Pat?' Martin asks, standing on a tree stump to get a better view. 'You hear the call?' She drifts towards us and Stephen runs his tongue over his top lip. 'Quite a young bird,' he says as she cuts back on the warm breeze, and then he turns to me and raises his hand like a claw. Blinking brightly in the light, he tells me some of the birds he's known have been 'absolutely intense. As soon as you look away, they'll fly down and crack you.' When Martin was last up here, the male bird was absent and Stephen explains that they'll often hunt up to 5 miles away from the nest. 'But he may have actually gone now,' he adds, 'because when the chicks get to between fifteen and twenty days old, the female will start hunting.'

Martin steps clumsily off the tree trunk and wanders on ahead, leading us through the bracken towards a rowan growing sideways out of the hill. As we walk and talk stops, the shrill song of siskins fills the silence and the harrier's call grows more and more agitated. 'The male,' Stephen calls from behind us, like a small boy watching a train go by. 'The male's just shown up.' We all turn and look towards the horizon where a grey raptor, black tips on outstretched wings, hovers above the bracken. 'The males are completely fantastic,' Martin turns and says to me. 'It's just, it's really, it's just a hell of a thing.' For a moment words fail him, and then he shakes his head with his hands on his belly and thoughts tumbling off his tongue. 'That display. It's just they're up there, in spring, one minute, then they're spiralling, sky dancing, spiralling down. The next minute they'll be just inches above the ground.' Gavin stands beside me. His face is almost expressionless but he stares with keen blue eyes. 'Acrobats,' he says, in a quiet, gravelly tone. 'They're acrobats. I took a video and I would have said it was a summersault but when you watched it back it was definitely a cartwheel.' Ahead of us, the female's song changes. Twice she whistles a sharp food call, and when I turn to look, the male has gone.

Hands brushing the waist-high heather, Stephen walks on ahead, talking as he goes. 'Gav's right, they just grab the air and spin on it but they're semi-colonial as well, which is quite special. See, if you've got two nests in a kilometre, another pair will show up and get in between.' As Martin sees it, their semi-colonial nesting is one of their biggest issues. 'Gamekeepers are terrified that if they get one, they'll soon have ten.' For the last five years, in the hope of allaying keepers' concerns, Natural England has been operating a brood management programme. The initiative allows grouse moor owners with more than one hen harrier nest in 6 square miles to have a brood relocated to somewhere where grouse stocks won't be threatened. In terms of fostering support from estates, Stephen tells me 'it's just been a revelation', but later when I am standing alone with Gavin he admits that 75 per cent of Natural England employees are opposed to it. 'Brood meddling they call it,' he says, shaking his head, 'but I'm not sure they know the ins and outs.' Kielder differs from a managed grouse moor in almost every way. Crouching down to run his fingers through the purple and green, Martin tells me that some moors, in order to encourage green regenerating shoots for grouse to feed on, are so burned 'there's just nae heather for harriers to nest in.' Conversely, on Kielder, the nesting habitat is almost perfect. 'We do have more foxes than gamekeepers would ever tolerate,' Martin continues. 'We had a nest down in the bottom fail this year to a fox but the point is, when you've got vast tracts of habitat, the harriers will cope. It's just natural predation. Foxes are meant to eat as well.

The trunk of the rowan grows one way and the branches grow the other as though the tree is losing a battle with the wind. Beneath it, Stephen stops and takes his bag off his back. 'At about twenty-eight days, the chicks can fly,' he says, as he unclips a large net. 'They'll almost be that now and it's a normal reaction for them to spring, so you've got to be ready.' Earlier in the month, I had been due to join Stephen at a nest nearby, but the chicks were killed by a goshawk before we could get there to tag them. 'One of those males had already fledged,' he tells me as he sets off across the hill. 'We'd seen him flying out and he'd just come back to visit the nest, one of those things.' Watching his feet as he goes, Stephen wades through the heather, holding the net out in front of him. 'Are we in the ballpark here?' he shouts down to Martin. 'We are, very much so,' comes the reply. 'In the heather here there's a bit of down.' Martin holds a feather out in front of him and Stephen tells me to tread lightly. 'The chicks will often move away from the nest and just sit there.' Above us, 40 yards up, flying against the breeze in a panic, the mother's cry grows louder and louder, watching us as we hunt for her young. Gavin stands next to me in silence, looking up. 'She's basically telling her chicks to keep their heads down,' he says after a while, his gaze following the bird. 'Even when you're not here she'll be chittering, but it's a different contact call just to let the chicks know everything's fine. That's not what she's doing now, though.' Below us, his hand held to his temple to block the glare of the sun, Martin is apologising again. 'It was definitely on a line between that rowan and that ridge.' Stephen tells him it's nae bother and turns to walk back up the hill towards us.

On the other side of the sweeping valley, where one country becomes another, a brashy plot of recently deforested ground shines brittle grey in the sun. Until the early twentieth century it was all owned by the Duke of Northumberland, and before that Kielder ran through the heart of the ancient kingdom of Northumbria. In 1932, in order to pay death duties, the Duke sold the estate to the government and much of what had been one of the country's most productive grouse moors was planted up to create Britain's largest man-made forest. It wasn't until almost seven decades later that hen harriers returned to nest at Kielder. Since then, a number have disappeared, fate either 'unknown' or 'mysterious', depending on who you talk to, but by 2017 the area had become one of the hen harrier's most consistent breeding sites in England.

When I turn round, Stephen has already swung his net into the heather and is crouched down over the chicks 60 yards above me. 'Got it, Martin,' he calls, then he counts them out loud and gets to four. 'Do watch your feet,' Gavin says in a not fucking around voice as we walk up the hill towards the nest. When we get there,

Stephen is clutching two young chicks against his chest. Their bright yellow feet and needle-like claws are held between his fingers and they stare up at me, soft brown eyes and black-tipped beaks agape. The brood size is just below average. In recent years, Stephen has seen as many as seven young and he thinks it's because they've started 'to compensate in terms of evolution for high losses'. As he looks down at the chicks, they tilt their heads back. Not with a look of fear but as though they're resigned to their fate, and all the while their mother drifts overhead. 'What we'll do,' Stephen says, 'is take this chap away from the nest a bit just to see if he's the best to tag.' He passes the bird in his right hand to Martin, whose t-shirt is dark with sweat, and then he takes ten steps and sits in the heather. 'You see that?' Stephen asks, holding the bird up to me. 'That down on his forehead, above the eyes? When that disappears he's ready to fly. He'll be our best option.' Behind us, the other young bird has settled in Martin's hand and is looking down at its brother, tongue half pink, half blue, hanging over the bottom of its beak.

Rummaging in his bag with one hand and holding the chick with the other, Stephen tells me that you 'sort of have to watch yourself. If you take your eyes off them, they'll hit bone instantly. Small but they're feisty as hell, tiny feet like needles.' I sit across from him and he glances admiringly at the bird then places it down in the heather between us. Some years ago, he took a vet out with him who measured a chick's temperature when he was handling it. When it was in his lap it kept on climbing and climbing, but on the floor it stayed stable. From a padded pack, Stephen pulls out a satellite tag attached to a harness. 'All the preparation is done in the house,' he explains, 'If I had to build the harness out here it would take too long. The first critical measurement is the body weight because the tag itself can't be more than 3 per cent of that.' Stephen smiles when I ask him how long the technology has been around. 'Only since 2007 actually,' he replies. 'The technology up until that point was too heavy.' He looks up, laughs, and says, 'It's funny really, half the time I can't remember where my car keys have got to but I tagged that first hen harrier. I was the first person in the world to do it and I can perfectly remember its ID number and ring number and everything.'

Martin appears at my shoulder, holding a small dead bird. I place my hand out and he drops it into my palm. The soft feathered body is still warm and there's a dark gouge through its skull. 'Juvenile pipit,' he says, looking at the blood on his fingers. 'Typical find from a harrier nest. That's what they'll be living on here. In the early spring, it would have been siskins, redpolls and chaffinches. This year there's no voles whatsoever and the grouse numbers are terrible.' Stephen shouts across to Gavin, 'Come give us a hand with the old gluing and stuff, mate.' I pass the little bird to Martin and he takes it back to the nest for the chicks. 'I think the male brought that in,' Stephen says when Martin returns. 'He did come over and he'll have probably just dropped it into the nest.' Picking the hen harrier chick up and putting it on his lap, Stephen starts stitching teflon braid to its harness. 'The bird's gonna wear this for life,' he tells me, as he works away with the needle, 'so it's got to be bang on the money. If it falls off and someone sees it, my name is mud big time then.'

When he's finished and the glue's been applied, he passes the bird to Martin who runs his pinky finger round the inside of the harness then holds the bird down to show me there's enough room for it to grow. Currently, the chick weighs just a couple of hundred grams, but in a year's time it will weigh up to 400, meaning it will be just slightly smaller, as an adult male, than the average adult female. 'Right, I want both of those numbers,' Stephen says, taking

out a notepad. 'The ring, down the leg,' Martin replies, 'is BA45217.'
Stephen pencils it down, 'and the tag ID?' Martin peers down at
the bird's leg then shakes his head. 'Can't bloody read that,' he
replies. Turning to me, he tells me to put my hands out. 'You need
to hold him secure, one hand on his legs so he doesn't claw you.'

When he passes me the hated creature, I hold it to my ribs and
it turns its head to look at me. It smells musky and sweet and
somehow it doesn't seem right. It's as though two worlds have
collided and something's been broken. I run my thumbs across its
dark mottled back and then I feel its heart starting to beat beneath
my fingers and I wonder whether human hands will hold it again
at some point, fate unknown. Stephen nods when I say it's a shame
really that we have to tag them at all and tells me that it is, in a
way, but as well as being able to track them acting as a disincentive
for people persecuting them, he thinks it has also granted us a new
way of seeing. 'What we've managed to do is just to have a look
at the landscape through the harriers' eyes. We used to think they're
upland birds that migrate to the coast in the winter, but it's the
birdwatchers who go to the coast in the winter. We've been able to
show that many of the harriers stay in the uplands year-round.' I
lift the bird up into the light to read the number on its leg. 'What
shall we call it?' Martin asks, as I pass the bird back to him. For
a moment I think and then suggest we should probably call it
Colin, really. Stephen shakes his head then gestures across the hill
and tells me we can't, there's already a Colin out there, and 'he's
thriving'.

Martin passes the bird back to Stephen for one final check and
he studies it with an intense silence. Above us, the mother's chittering has grown fainter. She has resigned herself, it seems, to
whatever's happened. I glance up at the bird and then ask whether
they think grouse shooting and hen harriers can coexist. Stephen

sucks the air between his teeth and smiles. 'That is the question, isn't it, but I think they can. It's just about people taking a step back. If a moor has 650 pairs of grouse, people might need to accept they'll end up with 600. That's all a harrier will take.' Gavin looks like he's about to say something and then Martin cuts in. 'We just need to take big numbers out of it. Shooting can be great or it can be shite, but that's nothing to do with how many grouse there are or hen harriers.' Martin also wonders if half the people involved appreciate that when raptors do well, it's often because of how much food shooting has provided. 'Part of the problem, to be honest, is that with gamebirds you produce a super abundance of perfect prey.' Colour flushes his cheeks and he laughs, 'Then they wonder why predators are doing so well.' I ask Martin about all those who would say they make their margin on those extra fifty birds and he shakes his head and tells me he's never understood it. 'People should pay a flat fee to go up on the moor for a good day, not per bird in advance.' Martin wipes his brow then says it's just not how the natural world works. 'It can't be the same every year. This year, everyone was excited in early spring because it was dry and there seemed to be a lot of grouse and then the cold came and it all went tits up.'

Stephen holds the bird out for Gavin and he scoops it up, then wanders back to the nest. I walk with him and as we go he tells me he's learned more about hen harriers from Stephen and Martin, out on the moors, than he thinks you could ever read. 'I was working before this as a foster carer,' he says as he places the bird down, 'and I just kept on observing nests and giving Stephen information and eventually I think he must have just seen a bit of potential in me.' He smiles as the bird runs through the heather, on unsteady legs, before hunkering down back in the nest with its siblings. 'You see the mother's still got brown eyes,' he says, pointing up at her.

'That's one of the ways you can tell she's young. In time, they'll turn yellow.'

Deep red on dark grey, six spots of blood lie on the tarmac beneath the stoat's brown snout. Its back is all bent and I expect it to straighten when I pick it up by the tail, but its body is stiff and it hangs curled and twisted in the air. Other than the blood and a few drops of pale-yellow urine on its white belly, the creature is almost perfect and I place it on top of the stone dyke on the verge to be eaten by something in peace.

'Not squashed?' Lindsay Waddell shouts through from the kitchen, when I arrive at his cottage in the pale afternoon and he goes to make the tea. 'Didn't seem to be,' I reply. 'Very bonny things,' he says over the noise of the kettle. 'Now tell me, do you take sugar?'

Lindsay was just 16 when he started his first keepering job on an estate in the Sidlaw Hills, which run between Perthshire and Angus. 'Before that I was brought up by an uncle,' he tells me, when he returns from the kitchen, 'but he was just like a dad to me really.' Whenever they were on the tractor together, out in the fields, Lindsay remembers that his uncle would always stop if they came upon a bird's nest. 'No matter what they were, he'd put the eggs in his cap and when we came back down the field he'd stop and then he'd just literally . . .' Lindsay puts one hand out in front of him and rubs his palm with his fingertips as though gently scraping out a nest in the earth, '. . . he'd put the eggs back in and we'd watch quietly, and a while later the mother would return. He was a farmer, my uncle, but there was a generation of keepers like that, men who were great naturalists, truly great men.' Lindsay sits up straight in his chair, looks to the kitchen door, grunts in thought then disappears for a moment before coming back with a tin of ginger biscuits.

He dips one in his tea, then gestures out beyond the window. 'These guys now, even the keepers up here in Teesdale, there are exceptions but they really are exceptions, most of them can't tell you many of the birds.'

After those three years in Scotland, Lindsay moved south to the Barningham Estate in Yorkshire. He smiles as he tells me that admittedly when he saw a hen harrier there, one autumn evening in the very early seventies, he stood and watched it for a while in captivated confusion. 'They were gone where I grew up in the Angus Glens and I'll be completely honest, never having seen one, I wasn't very sure what this bird was, but a lot of those first encounters really stick in your mind. They really do.' In every window around the room, dozens of frenzied flies are knocking against the glass. 'These bluebottles are enjoying it in here today,' Lindsay says, turning to watch them. 'They always come in when the weather's muggy and close.' In the distance, up over Fendrith Hill, the sky has darkened to a warm leaden grey and the rain is rolling in.

Since Lindsay first drove grouse to waiting guns, he has seen huge cultural changes on the moors. Back when he was young, despite feeling there was far more love for wildlife among keepers, he admits there was little tolerance for the hen harrier. 'If we'd carried on as they did at one time,' he tells me, while topping up my tea, 'shooting would have ceased to exist. We couldn't have carried on like that. Political pressure, in the last decade, has really ramped up.' While attitudes back then may have been entrenched, Lindsay thinks it was in the 1990s that the rot really set in on estates across the uplands. 'There were people who were just upping the ante all the time,' he says, shaking his head, 'at the very point when they should have been easing off.' For over a century, keepers have known that grouse are susceptible to a parasitic worm that lives in their gut. Down the years, its presence caused dramatic cyclical peaks and troughs in grouse populations, which meant there were often insufficient numbers for shooting to take place. But in the 1980s, a lucrative dawn broke across Britain's moors when it was discovered that leaving out mounds of medicated grit could reduce worm burdens by up to 40 per cent. At last, artificially high numbers could be maintained without worrying about nature attempting to bite back in order to restore some sort of balance.

Lindsay's cheeks have been burned red from a lifetime of facing the wind, and when he laughs, deep craggy lines sink across his forehead. 'The whole grouse industry made a real mess of this,' he tells me. 'Man's greed took over. We were given something which would have allowed for a bit of consistency but people wanted more and more.' One of the most regrettable effects as he sees it is that it 'ruined the sale of grouse as a product'. It went from being a real luxury that every smart London restaurant wanted to being something that estates had so much of they could hardly give the birds away. Lindsay thinks it madness that keepers continued persecuting hen harriers when there was no real need to. 'We had so many grouse, we couldn't shoot them. Why kill something you don't need to kill? You simply don't have to kill hen harriers when you've got that many grouse. We were awash with them.'

When it flashes, it happens so quickly that I look up for a moment, wondering if it happened at all, but then the thunder breaks, flowing down the valley in full fury, and heavy rain starts falling. The consistency that the grit ensured meant that agents started to appear, promising grouse moor owners they could turn their estates into highly productive and profitable businesses. 'Those men were bad news for shooting,' Lindsay tells me, in a voice that runs from Angus to Cumbria and back again as many as two or three times in a sentence. 'The owners initially thought they were a great thing and they were the very worst.' He sits back in his chair and raises his hands to his mouth in thought before starting again slowly. 'They brought in methods and practices – they just paid no heed to the law. They were blatant about it. They did it openly, as much as to say, we're above all this, you try and catch us.'

Three mallard fly high into the wind above us then drop down and circle twice above Lindsay's cottage. He turns to watch them, telling me three times, in three slightly different ways, that the ducks have done well this year in spite of everything. When they land 100 yards out, in a sheep field, he sits in silence and lifts his head to see them feed. 'When I retired, here on Raby', he tells me, turning back, 'after thirty years as headkeeper, I asked my old boss if could stay on the estate because I wanted to be in Teesdale and they own the vast majority of the houses. I wanted to be here because I've got snipe drumming away. I've got curlew, I've got lapwing, I've got redshank, I've got oystercatcher. I've got everything. I've got it all.'

During those years of manufactured plenty, Lindsay thinks there was no doubt that young men were let down badly by people who had a responsibility to them. 'Not the slightest doubt about that,' he says, shaking his head. 'I mean, if you didn't come up to scratch you were gone.' He stops a second as though listening to the rain coming down on the glass above us and then tells me about some of his young underkeepers who he walked the hills with, teaching them to care for it all. 'What went on is wrong,' he continues with a shrug, 'but it happened. It's happening less but it happened.' As well as young men, Lindsay thinks it's important to understand that older keepers too became victims of a rotten system. 'My friend Brian, an absolutely fascinating character and a headkeeper here in England, he was finished. He was finished because he was a traditionalist. He wouldn't have wanted to do the things those men wanted him to do.' Lindsay laughs when I ask where he ended up.

'Out on the Mulls, up on the West Coast of Scotland, he got a job there caring for a pair of eagles.'

Just a couple of weeks before I set out to see Lindsay, news broke that it had been the most successful season for hen harrier chicks fledging in twenty years, and out of seventy in total, nineteen were on moors managed for grouse shooting. What nobody knows though is whether they'll be left alone. 'You know,' Lindsay says, 'I think they will be. They've got to be. These grouse lads, I feel, have got the message. They've got to tolerate these birds because if they don't, if they were to carry on as they were, they're not going to exist for much longer.'

Up on the mantelpiece, a jam jar of brushes sits next to an old clock. 'Do you paint?' I ask. Lindsay stands up, walks to the corner of the room, and lifts three sticks out of an umbrella stand. 'I used to,' he says as he passes them to me, 'but only now when I'm doing these.' In fine detail, on the tup horn handles, he has carved a snipe's head, a grouse and a trout. 'I just do them because I enjoy it,' he tells me, 'but I'm not happy with the trout. The expression is a little strange.' I look at it closely. The fish is frowning. 'You know, it's not just grouse and hen harriers,' Lindsay says, as I run my fingers across the trout's sad human eyes. 'I've had things in my palm, as you will have done, salmon parr, lapwing chicks, plover chicks, and you look at them.' He holds his hand out in front of him and stares down at it. 'I don't know. I honestly don't know what it is. I would tell you if I could, but if you don't get some deep appreciation from these creatures you shouldn't be working in the natural world.'

The stoat on the dyke has gone and rain runs across the road, washing away its blood. Cumbria is still full of rabbits and I swerve all over the road counting them along the moorland fringe, tucked up and grazing where grass becomes heather. The road goes right, cutting up to St John's Chapel, but I turn sharply left and wind my way up over Wearhead and across the old lead mines – Alston, 10 miles. At the side of the road, over the top, two grouse sit on an old wooden rail, dark silhouettes in the mist and damp. Garrigill is shut up when I get there. The first leaves of autumn lie scattered across the main street. The grand lady's roses have long lost their petals. No lights on in the village shop and the George and Dragon is closed.

The lapwing act

I only knew them gone

When, out of a sad winter, one returned.
I heard the high mocked cry, 'Pee-wit,' so long

Cut dead. I watched it buckle from vast air
to lure hawks from its chicks. That time had gone.

Alison Brackenbury, 'Lapwings',
Gallop: Selected Poems, 2019

Twenty-two feet across, the big wheel turned 75 times a minute, 40 ropes and 40 tonnes of piston and plate. Scarred-up lungs from the cotton dust and slowly deafened by the noise, the workers wove from dawn until the sun went down over Rochdale.

Beneath the footbridge, the canal that once provided water for Arrow Mill's great steam engine curves away in front of me. Yesterday's snow, illuminated by the lights on the towpath, lies thinly blown across the frozen surface. On a flat-roofed warehouse, a robin starts singing its inquisitive song, but the bird is hidden in the blackness behind the coiled razor wire. Further along, a door graffitied with a silver cock and balls opens and a man in a blue hat and blue overalls steps out, leans on an air vent beneath an orange bulkhead and puts a cigarette in his mouth. He's most of the way through before he seems to notice me, and when he looks in my direction I turn away and cast my eyes over the ledges and the red brick chimney rising up against the dark cloudless sky. I take in every bit and then find, on glancing back, that he is looking up too, as though trying to decipher my secret, but there's nothing to see anymore. The lapwings stopped coming some time ago. He flicks the cigarette onto the gritted tarmac and the balls swing shut.

Nobody is quite sure when lapwings roosting on roofs in the North West really began, but if you ask around enough, you'll be told stories of them swirling over old industrial buildings as long ago as the 1970s. Initially, it seemed to be unique to Rochdale, and Arrow Mill was probably the first roof that the lapwings ever took to, but in the 1980s and 90s the birds started to be observed more widely, with a list that eventually ran to 16 sites, including bakeries and factories, from Wigan to Monkey Town. In more abundant times past, there were rare reports of flocks of up to 600, but it's been decades now since even half that number were seen flighting in across the Manchester sky after a hard night's feeding.

It isn't totally clear why the flat roofs have proved to be such attractive places. Some reckon the lapwings like the warmth, but in truth they don't have much choice. The fields where they once roosted in their thousands have been cut up and turned into housing or have had tarmac poured across them to provide parking for industrial estates and out-of-town garden centres. We claimed the places that were theirs and they were forced to take refuge on what we left behind. Once, just over twenty years ago, a pair stayed on a Rochdale roof the whole year round and reared young. The chicks apparently scraped by on invertebrates clogged up in a drain, but for the most part the flocks head off in early March in search of undrained open country, where they make their nests.

In the spring of 1908, during Britain's coldest ever April, just a year after the first brick was laid, Arrow Mill was declared open. At the ceremony, Mrs Waller, the wife of the Chairman, christened the steam engine 'Reliance'. The mill was a fine building then and it still looms grandly in the darkness, but as the sun comes up and cold blue seeps into the sky, I see that the windows are smashed and buddleia is pushing its way out of the cracks in the bricks. Up the lane, in the terraced houses, a radio plays and lights are starting to shine between the gaps in the curtains, people climbing out of bed for another Monday.

The air is sweetening with the taste of diesel and I'm trying to work out where the fug is coming from when the cry of Canada geese sounds over Castleton High Street. The noise draws closer and closer but the birds don't appear until they are almost overhead, seven dark shapes above the willows, wings set, dropping onto the canal behind the footbridge. When they land, a rich cracking sound draws across the ice.

On Sherwood Street, in the shadow of the Mill's balustraded water tower, a red, white and blue sign on the Queensway Fryer promises a full English from 7 a.m., but a metal grill has been stuck across the door and there are plywood boards nailed to the windows.

A Mini tears up the outside lane, T reg and green, but the motorway, otherwise, is quiet for a mid-afternoon Monday. Kane Brides and I are standing on the flyover, huffing hot breath into our hands, coffees steaming on the rail. Knowing what he does now, he tells me it makes him feel a bit ashamed but he thinks that in truth it was what gave him a love of lapwings. 'Frank, my grandad, would take me down to a farm, only about three miles from here, and we'd flush the bird off its nest.' He shakes his head and pulls a pair of gloves on as he talks. 'The mother would swirl above us, calling in alarm, while grandad would show me the beautiful dark brown, dark greenish speckled eggs.'

When Frank was a child, before he left school at 14 to go down Pretoria Pit – where just a couple of decades earlier, 344 men lost their lives in one of Britain's worst mining disasters – most little boys collected eggs. The real prize, he told Kane, on one of their walks, was having a whole clutch – in the case of the lapwing, everyone wanted the full four. As a boy, Frank spent hours on the land that stretches out in front of us. It was all grass then and he knew every tussock and every glade, all the secret places where a bird might lay an egg. Almost half a century later, in the 1990s, when Kane was small, the fields were still much as his grandad had known them and Kane remembers 'goodly numbers of breeding birds', but in the space of just a few years almost all the lapwings went. Not driven away by excited little boys with boxes under their beds to fill but pushed out by all the new houses being thrown up along the side of the motorway. 'It was greenery right the way to that stadium,' Kane tells me, pointing into the distance. 'I was four when that was built, the home of the Bolton Wanderers.'

When Kane and Frank went out birdnesting, the old man would tell the boy that they were only looking. Taking, at that point, due to its formerly catastrophic impact on lapwing numbers, had been illegal for years. In the nineteenth century, droves of pickers poured over the countryside every spring in search of nests. Norfolk was particularly hot and in just one year a lone picker managed a bag of 2,000 eggs from an estate near Potter Heigham on the Broads. At a going London-market rate of 3 shillings a dozen, the haul would have sold for a similar sum to a farm labourer's annual income. Usually hard boiled and sometimes covered in gravy, the eggs were sold as 'plovers' eggs', owing to the birds' older name, the green plover, which derives from the Latin for rain, *pluvia*, on account of the belief that lapwings start to flock when the weather is on the turn.

By the 1880s, most of the countryside had been stripped bare as far up as Lincolnshire, and shipments of lapwing eggs had to be sent down from the north where the birds were holding on. Throughout the late nineteenth century and into the early twentieth, naturalists voiced concerns that unless something was done, England was set to lose one of its most beloved birds. At just after eleven on a rainy early-March night in 1928, the Protection of Lapwings Bill was read out in the House of Commons. The minutes record that talk of soldiers' pensions had dragged on all evening and the members were sleepy, but a change of step, from war to peace, brought them back to life. The bill proposed that it should be prohibited to pick eggs, to be sold for human consumption, between 14 March and 11 August. The hope was that this would end the lucrative exploits of commercial operators while still allowing the odd one to be taken for a person's own dinner, as well as leaving little boys like Frank Brides to their collecting. It seemingly didn't occur to the MPs that little girls might like birdnesting too.

Sir Joynson-Hicks, the puritanical Home Secretary who spent most of the 1920s in a flap about nightclubs and everything he suspected happened in them, read out the bill. He reminded his comrades that lapwings, as birds that eat crop-devouring crane fly larvae, are great friends of the nation's farmers, but added that what with plovers' eggs being so succulent he felt terribly sorry for 'the epicures'. At well past dinner time, when Joynson-Hicks was struggling to keep members on track as they coughed up cherished memories of boyhood birdnesting, Colonel Applin, the Tory MP for Enfield, made his contribution by asking why they're called lapwings. It seemed, he thought, 'quite a country name for the green plover'. Joynson-Hicks replied that the birds are known as lapwings throughout the whole country and assured him 'it is a good old English name'. He was right at least about the origin. In the fourteenth century, they were known as *lappewinkes* because of their jaunty flight: wince meaning to waver and lap being to leap. The Glasgow MP, George Hardie, Keir's younger brother, wondered 'why not peewits' – and beyond Britain's cities, particularly in Scotland and the north, on account of their piercing 'peewit-peewit' cry, they still often are. Ayrshire Minister, James Barr, the Member for Coatbridge, was keen to add that he knew them by their less common name, the green-crested lapwing. It is possible he had come across the term in Robert Burns' *Sweet Afton*, in which the poet asks the screaming 'green-crested lapwing' to let 'sweet Mary' sleep. The last word went to Lieutenant Commander Kenworthy. The prolific author and radical liberal was worried that unless they added some definitions to the bill, 'country lads' from his Hull constituency might set out in search of lapwing nests, innocently thinking they were only picking 'peewit eggs'.

Twenty-six years later, in 1954 when the Wild Birds' Protection Bill was debated, a total ban on collecting lapwing eggs was narrowly fought off by Lady Tweedsmuir, daughter-in-law of the novelist, John Buchan. Displaying a degree of knowledge that would be largely absent among parliamentarians today, Tweedsmuir argued that lapwings can lay up to five clutches during the breeding season and noted that the first one seldom survives due to harsh conditions or being smashed up beneath the plough. Tweedsmuir's belief was that if the first clutch was taken, it would encourage the lapwings to lay again, when spring had come and the weather was kinder. But by 1965, it was generally accepted that the experiment had been a failure. A proposal to extend the protected period was raised in the House of Lords and was enacted two years later.

Three hundred yards down the motorway, behind a wooden fence beyond the hard shoulder, two herring gulls float over the green warehouse roof of a recycling plant. 'My lapwings were up there all last week,' Kane tells me, 'and I was hoping this cold weather was going to anchor them for us, but they've maybe gone west to Ireland.' He puts his binoculars down. 'That's their favourite spot. There by the skip hire building.' Kane can't remember where they were headed but they were driving north and he was 14 at the time. It was his dad who spotted them first, all flocked up and displaying over the motorway. My coffee has gone cold and I pour it into the drain by our feet. 'I've just always loved a proper performance,' he continues, 'whether it's white-fronted geese or lapwings, ever since I was small.' He smiles and shrugs. On the way back from wherever it was they'd been, Kane had his nose pushed to the window, hoping to see the birds up in the air, but they'd settled on the roof, tucked up on the green steel. Ever since, whenever he's driving back home from Slimbridge, on the banks of the Severn in Gloucestershire, where he carries out research into wildfowl and wading birds, he comes up onto the flyover just to check if his birds are still there. Lately, he hasn't been able to get back to Slimbridge and he thinks it might be some time yet. During his months at home, he's seen more of the lapwings but the flock is smaller than he remembers.

Recently, Kane has been reading up on the impact of foxes and badgers on ground-nesting birds and he suspects that part of the appeal of the roofs is that they give the lapwings some respite, but predator pressure and houses being thrown up where they once roosted are far from the only problems they face. Following the ban, in the 1920s, on picking eggs commercially, lapwing numbers started to stabilise, but the positive impact was subsequently swept away in a century of agricultural intensification. 'It's a whole load of things,' Kane tells me solemnly, 'so at one time, up until about the 1970s and '80s really, farmers would be drilling crops in spring, but now we've got so many winter cereals. The crops are basically sown in autumn then harvested the following summer.' He lifts his binoculars to watch a heron cut across the motorway and continues talking as he tracks it, 'so when the lapwings are starting to look for breeding spots the crops are now too tall for them. They like swards of between 5 and 15 centimetres. Anything higher than that and they're really not happy birds.'

The boom in winter cereals, as part of the story of Britain's agricultural intensification, is the chapter that followed the dawn of our agro-chemical age. While lapwings were lauded in parliamentary debates for being the farmers' friend on account of their hunger for crane fly larvae, they were no match for synthetic insecticides like the now-banned DDT, which fulfilled the lapwings' pest-control function with clinical efficacy and left them to starve.

The road beneath the flyover is becoming busier and the sky out to the east, over Winter Hill, is darkening with grey snow clouds. On the walk back to the services, Kane stops to glance through his binoculars, but the dozen shapes gathered in the middle of the field beyond the petrol pumps are only magpies. The following day, Kane must help his grandmother pack her things. She has lived in the same house for 46 years but it is time now for her to go to a care home. 'Suffering from dementia, bless her,' he tells me fondly, 'but we were having a cup of tea yesterday and she was talking about the mills. She worked in the cotton mill in the same town where grandad was a miner. She said she just loved getting up early in the morning and going off to work. She was reciting it all as though it was last week. It was really quite beautiful actually.'

'There was Woman and Wife, there was Upside-Down Face and there was Handbag. Upside-Down Face had a big beard and was bald on top, Woman and Wife were two blokes who went everywhere together, and Handbag was called Handbag because he turned up at the docks every day holding one but nobody ever knew what was in it or why.' James Walsh and I are sitting on the top tubes of our bikes, looking at a grainy portrait of his father through a porthole in a weathered steel sheet. James reckons the two cargo cranes, which rose up above Ontario Basin, were a better memorial to those who made Manchester. But in 2011, as the locals looked on, the council smashed them up, and now twelve structures, created by the sculptor, Stephen Broadbent, in the shape of dock workers' union cards – each of them featuring a picture of a man who would have carried one – is the only scrap of Salford's past on the quayside. 'Basically, what I remember were these proper hardcore blokes. I never knew their real names and me and my sister would spend all our time down here running in and out of Portakabins just having a laugh.' James stops for a moment and sucks at his juice then wipes his mouth and carries on. 'I was born in 1976 and the docks stopped in 1982, so basically all the old like proper offices were already closed down.'

Bill Walsh, greased side-parting, deep-set eyes, and a docker like his father and grandfather before him, is a young man in the porthole picture but he was in his fifties when his own boy was born and he was working, at that point, as a welfare officer. 'Thatcher and Murdoch and all that really dumbed down what people think of the working classes,' James tells me, 'but the dockers would pull a fifty-hour week then they'd be doing rumba on a Thursday night. Dad put on all sorts of South American ballroom classes.' For those who didn't dance there were Spanish lessons at the pub, standing room only for the many dockers who had links to the *socialistas*, newly free of Franco. Just a few years after the ships stopped coming in, James remembers being a young twitcher and watching his dad's old docks bursting into life, 'in like the mid-80s, there were pochard feeding at night, there were skylarks, and it were full of lapwings.'

Behind us, beyond the towpath, a lingering mid-morning mist rises over the River Irwell and drifts out through MediaCity's shiny steel avenues. James says he thinks blokes like Handbag and his father would just laugh if they could see what a yuppie paradise it's become, but to him, more than anything, it's a missed opportunity. 'The whole docks could be a world-class nature reserve and every building, all the flats, could have a green roof for birds.' James doesn't think that redevelopment and nature are totally incompatible but that 'capitalism just needs to be a bit more imaginative'. A couple of years ago, he managed to get a meeting with the Mayor of Salford where he shared a plan to launch some electric boats shaped like swans that could go up and down the river, showing people the last of the lapwings. All he got, though, was a smile and a 'good luck, lad'.

James clicks down through his gears as though we're going to set off but then leans his bike on a graffitied bench, swings his rucksack round and gets his camera out. He takes it everywhere and he wants another photograph of his father. 'The sculpture's called Casuals,' he tells me, as he stands in front of Bill, 'because of how casual it was. That's what the workers were called. People would walk down to the docks like five times a day and not get work.' Although he reckons being an 'ecologist is a bit of a different setup', James says he sometimes feels like a modern-day docker. 'There's not enough jobs about and employers are only interested if you'll work for next to nothing.' To Bill's left, on a steel sheet leaning the other way, pictured behind another porthole, there are three laughing young men sitting on wooden pallets, shirt sleeves rolled up, empty tea cups and cigarettes between their fingers. 'Me and my mate always work from home,' James tells me, looking at the happy faces, 'and we just carry on like that all day.'

As we pedal west, past the sewage works, James shouts over his shoulder. It's something about making the lapwing the official bird of Trafford, but I can't hear much over the roar of two lorries stuck behind us. When he pulls his brakes on and bumps up onto the pavement, I swerve to avoid going into the back of him, and by the time I've righted myself he's looking at an open gate in disbelief. 'When I was like a proper mad twitcher I used to come here all the time, but then for ages they started locking me out.' Three builders talking to a man with a clipboard eye us strangely as we wheel our way through the small car park towards where it drops down to the river. 'Down there,' James says, resting his bike against the railings, 'you are literally seeing the last stronghold of the lapwings on the quays. They've just been driven out by development. Totally driven out.' James reaches into his bag for his binoculars and talks slowly as he leans on the rail in front of us, scouring the landing bay on the north bank. 'On that bit of grass, see, they nest every year. It's probably about six years ago since I got in and last saw them, but they're here every spring. The sewage works absolutely reek in the summer but it's properly good for the lapwings to feed on.'

His voice rises with excitement as he tells me to get my 'bins' out. 'Look across at the oystercatchers. Then a bit to the right of the oystercatchers. You've got that sort of holly . . . near that holly.'

At first, I can't make it out and it seems James doesn't want to say the word in case it spoils me seeing it for myself. A moment later, all crested head and dark green back, I see it sticking up out of a scrape in the moss. I can't find the words to make it sound as though the encounter is transformative. The lapwing is sitting quietly, only half visible. I feel awkward and I pretend to be transfixed, but James starts shouting. 'Up to the right, there, up to the right!' I put my binoculars down and turn to look along the river. In the sky, a male lapwing tumbles towards the water, dropping as though dead before stretching his wings out, steadying himself, and rising again, his sweet peewit cry cutting over the traffic.

James had been birdwatching for about two years when he first saw a lapwing. He was six at the time. He knows because he drew it with his colouring pencils. First of all, he tells me, it's a spiritual thing. 'You look at birds and they can fly.' Six feet below us, a Canada goose is drifting past. 'That Canada, you're looking at something that if it fancies it can just go and fly halfway round the world.' James thinks there's a feeling at the BBC that when they launched themselves on Salford, they brought something cosmopolitan to the dirty old town, but he reckons it's nonsense. 'One thing I've been trying to tell people is that Salford Docks used to have all these global connections through trade, people would dock here for like two nights from all over the world, and now the birds is all what's left of that. We've got a sand martin colony. Obviously they're coming from Africa, pochard goes to Eastern Europe, and even the lapwings can go to Ireland.'

Up the hill, over the railway line, we cycle shoulder to shoulder and James tells me that the only other thing he's ever really been into was raving. 'It was like five nights a week at one point. I was 10 stone lighter then. First record to the last record, I'd be on the dance floor.' We cross over a roundabout towards Eccles and I ask whether he started birding more when he stopped partying. 'I used to do both,' he replies. 'Even this New Year, me and my mate managed to get our hands on some acid for the first time in ages. Like obviously house parties have become the thing. You just got some DJ in his garage broadcasting to the world. We just got messy all night till like four in the morning and then went straight out birding.' On our right, we pass a theatre where James once saw a play about the lives of the Salford dockers. 'Really immersive,' he tells me, catching his breath. 'Birding on acid must have been sort of immersive too,' I reply. James nods. 'Aye, it was certainly interesting.'

On a small screen fixed to the wall above the pickled vegetables, a white maggot contracts and opens in a forward moving motion across a grey table beside an old book. The music is tender and lonely. The video cuts to twin baby girls lying next to each other on a blue bedspread. Then it fades to a beach – a man in a long black coat walks along the sand and a gull flies over the sea. Then there's a dog, then chickens, then a party. Beautiful young people dance in a kitchen, bodies pressed together, then a lady in a bath eating an orange. Halfway down the aisle, an old man leaning on his trolley is watching too. When the video ends, he turns to look at me and smiles, then he shuffles away towards the cold meats. Out on the street, James is squatting on the pavement, trying to bend his derailleur back out so it stops rubbing against his spokes. 'I got you a Lion Bar as well, just to play it safe in case you don't like the other one. I don't know my way around the chocolate in Polish supermarkets.' He stands, wipes his hands down on his ripped grey tracksuit bottoms then opens the wrapper and eats. 'That is really nice, that. Do you want a bit? They do know how to make chocolate, the Poles.' I shake my head.

Away from the river the sun has burned through and the day is cold. A lady walks past with a cat in a carrier then turns into the vet's, but otherwise, apart from the Zabka mini-market, Liverpool Road is quiet. 'I used to live just down there,' James tells me, pointing back the way we came. 'The landlord wasn't good, though.' I bite down into my Carpathian mountain cake and sweet cream oozes onto my fingers. 'What was wrong with him?' James replies between mouthfuls. 'Violent. Alcoholic. He looked like Mussolini if Mussolini had a younger brother. Me and my mate spent all week taking various items out the house in the middle of the night and then we did a runner. Haven't seen him since.' I finish my cake in silence, eyeing up everyone who passes to see how much of a resemblance they bear to the Italian fascist. As we get on our bikes, James says that to be honest it wasn't the best but they had fun all the same. 'In the 1950s, the tourism marketing board had these posters going on about "The Costa del Salford" so we'd do the Costa del Salford disco.' The number 67 bus pulls round in front of us and we pedal along in the warmth of the exhaust fumes. 'What we'd do is we'd lock the ginnel up so you could only get in through the houses, pour tonnes of building sand out, and then invite everyone round. We'd have a gazebo, sun loungers, cocktail glasses. Oh aye, me and my mate have a million tales from round here.' On our way out of town, beyond the retirement homes and semi-detached bungalows, we pass a man up a stepladder outside a flat-roofed church. He is reaching up and sanding down Jesus's legs. The winter weather has left him pale and flaky and he needs another coat.

Beneath a billowing sky, in an 1851 watercolour by David Cox, two women pad across Carrington Moss with bundles of faggots on their backs. Cox, as a young man, was apprenticed to a miniaturist and in later life he perfected the uncanny art of imbuing small characters, set in great landscapes, with a very precise and full sense of being. The two hunched figures look weary and are struggling across a wild sweep of heath that is about to change. Carrington Moss, up until the mid-nineteenth century, was a grouse moor, but as Manchester expanded, and the amount of sewage the city produced rose, the heather was reclaimed as a place for a daily dump of over 40 tonnes of night soil. By the 1930s, the English had taken to the flushing toilet, resulting in a great decline of shit needing to be dropped on the land. 'In them days, around the 1930s and '40s,' James tells me, as we pedal along a lane that cuts through the fields, 'there were like 200 pairs of lapwings here on the Mosses but there's only just remnants of that now.'

On a bridge that runs over the M6, James gets off his bike and leans on a padlocked gate. Fifty yards down, green metal barricades rise up 12 feet tall with a coil of razor wire running along the top. As he sees it, we are looking out across a battlefield. What quickly became clear, he reckons, is that the suits at IGas believed there'd be no opposition to them fracking the Mosses because they didn't think it was a place anyone actually cared about. Back then, in 2014, the green fortress beside the motorway was the IGas outpost and if they'd won, James believes there'd now be a fracking pad in every field where the lapwings still nest. Draining the last of his juice, he tells me that what he remembers more than anything is just how absolutely beautiful it was in the face of police brutality. 'There was so many gorgeous people. It was like something New Age. There was a guy called Phillardo Fumblefoot who came up from Portsmouth in a mad caravan, and he had a wife called Guinevere. They were druids and lovely people with it, and there was a wonderful woman called Vanga. She just got battered by the police. There were a lot of photos that went out of her getting battered.' All through the winter the protestors camped out in wigwams and would block the traffic every morning. James smiles as he remembers when he made dinner for everyone in his flat in Oldham on Christmas Day before ferrying it up to the Mosses with his mate.

Across the field, on a muddy bank, where the earth drops down to a stream, a lone lapwing is perched on the mud. Through my binoculars I can make out a tall crest on its head, and its tail is bobbing up and down. As I scan the stubble, trying to find the female that the male bird is displaying for, James says he sometimes thinks about what it would be like to come to the Mosses one spring and not see a lapwing. 'You'd be devastated. You'd be devastated if you came out here and never saw a lapwing, wouldn't you, absolutely devastated. I would feel I'd failed.'

We cross the bridge and James tells me that beyond fracking a lot of the land is earmarked for housing development, but while he was protesting he started to think there was maybe another way. 'I came up with this idea called Big Mooch on the Moss. We literally just spent all day mooching. You could do that every day. Could be like five quid, ten quid. This is where capitalism once again is missing out on loads of opportunities because they don't see the bigger picture.' A silver BMW rushes up behind us and I drop back to let it pass. The driver looks younger than me and a girl sits in the passenger seat. For a while we pedal in silence then James turns and shouts back, 'Thing is, we mooched once already. We got Bez from Happy Mondays involved.'

The sky is hardening across the Mosses, heavy grey clouding out the blue, and four lapwings criss-cross over each other, whiffling to the ground before riding up again over the winter stubble. One peels off and flies upwards to mob a magpie drifting overhead, but the three others continue as though performing just for us. An old woman on a mobility scooter, small body weighed down by blankets, wheels past then stops 10 yards further on to see what it is we're looking at. 'I mean, they're completely underrated, are lapwings,' James says as the lady heads off again. 'I think they should almost be our national bird.' The magpie turns back and the fourth lapwing rejoins the three. 'I just think they can link the urban and rural,' he continues. 'It can be so many different habitats. They want moorland and waste ground and farmland, but I've even seen them roosting on B&Q in Altrincham.'

A strong wind is coming across the flat fields and our voices are getting lost on the breeze. James wheezes as he says he keeps thinking he'll know when Manchester becomes a world-leading green city because there'll be an increase in lapwings. For a mile or two, I tuck in behind him as he battles the wind. Where the tarmac runs out to cinder, he stops and tells me there's a café in a Portakabin a bit further on, closed last time he was up, but it might have reopened. 'Have you got lights, by the way?' he asks, as we head up the track towards it. 'I can spend all day up here. I'll show you where the birding hide used to be, until some lads burned it down.'

Beneath the bridge, cutting through murky water lit bronze by the hot Friday sun, a bank of shingle heaped with bladderwrack runs out into the estuary. For almost 900 years there were ferries that crossed the Severn. As far back as 1131, the monks at Tintern Abbey started using a mile-long stretch between the village of Aust and the Beachley Peninsula, which down the years came to be known as the 'Old Passage'. In 1725, Daniel Defoe decided he'd go to Wales the long way round and carried on up to Gloucester where the river is narrower. The passage, he thought, looked 'ugly' and 'dangerous', while the boats seemed 'very mean'. As time passed, the ferries improved but demand also increased. By the 1950s, the boats could only take seventeen cars on deck and queues were growing ever longer. In 1966, the American photographer, Barry Feinstein, took a picture of Bob Dylan in the rain outside the shop at the Aust terminal. The picture was later used on the posters for Martin Scorsese's 2005 film, *No Direction Home*, which traces the singer's journey from folk troubadour to becoming the voice of his generation. In Feinstein's photograph, in the fog, you can just make out the newly built bridge. Dylan was one of the last people to catch a ferry across the Old Passage.

Windows down, I drive slowly up the outside lane, leaning out to try and see St Twrogg's Island, where a ruined medieval chapel sits on the rocks. It's thought a hermit once lived there, and after he was gone pilgrims started coming, but as the sea rose, it became harder and harder to reach and slowly the walls tumbled into the tide. A car rushes up the inside lane, the driver twisted round, one hand on the horn and two fat fingers thrust up at me, his eyes twitching as he shouts. From the passenger seat, a lady looks across, shaking her head slightly. When I shout back, he points up to the hard shoulder, gesturing for me to pull in where the bridge returns to the land, but his hands are twice the size of mine and the woman's face says he's fought before.

Thick dark hair cut short and large sloping shoulders beneath a check shirt, Charles Grisedale leans past me and plunges his fork into a burger. 'Grisedale wanted burgers,' Lizzie tells me, across the table, 'because you know we do our own beef.' Charles nods and mutters something about burgers being all they've got left and a bloody good product too. He chews for a moment then stops and holds his arm out in front of him, whereupon Lizzie picks up a tube of English mustard and presses it down on his palm. 'It was different then, you see,' he begins, placing his knife and fork down on his plate. 'There weren't so many paedophiles about and there were 500 lapwings in the parish flock.' When Charles was a boy, he remembers some men coming to the farm to lay drains in one of the fields. 'On the Sunday, after they were done, they asked my mother if they could go out for a shot. They wanted a rabbit and they said we'll take Charles with us. She said yes. As I say, it was different then.' All afternoon the men and little Charles walked back and forth across the 300 acres without seeing a thing. He remembers them saying that they'd have to shoot something and then in the top field behind the house they came upon the lapwings. 'There was one wanted to have a go and I said absolutely bloody not, so we counted them instead, and that's how I know there were exactly 500 in the parish.' Lizzie looks at Charles lovingly then glances at me and smiles.

Behind Charles's head, on the stone wall above the fireplace, copper pots hang on hooks next to an old milkmaid's yoke, and beneath them, its head turned towards us, a carved wooden lapwing cock perches on the dresser. 'Thirty thousand of them in Wales in the 1960s,' Charles says, pouring himself more beer, 'to just three hundred now. It's shocking and nobody really gives a damn. The old thing used to be if you got the habitat right, the job's right, but they haven't got a fucking hope in hell with Billy Brock everywhere with his rather fantastic nose.' Out in the evening sun, just below the window, a loud whine starts up. Charles peers down at his watch then says, 'He'll just have to wait. We're still eating.' With her fingers clasped to the table, Lizzie looks up at me and whispers, 'That's the peacock. He's come for his fruitcake.'

Back in 1999, when Charles was 40, he started to realise that every year fewer and fewer lapwings fledged at Cefngwyn. 'I lose track of time but I suppose that'll be 22 years now since I put the fence up around 240 acres. There's no bloody grant on this job. It's my hard-earned cash that paid for that fence.' By Charles's reckoning, now that so much habitat has been ruined, badgers could be the end of the lapwing. 'Something like 500,000 in England and Wales,' he says, while reaching for the potatoes, 'up almost 90 per cent since the 1980s.' Lizzie walks to the fridge and comes back with a loaf of fruitcake and a block of cheddar. 'If they'd just stopped people badger baiting,' she says as she sits down, 'and they let the farmer cull them now and again, things would have been a lot sweeter in the countryside. My father, in Lincolnshire, would just say we got a pig in, Liz, and he'd quietly go get rid.' For a while, we eat in silence and then Lizzie smiles. 'You know what, you'd have liked my father. He'd have sat you down and he'd have said now then, tell me everything you've seen and he'd tell you everything he'd seen. He knew all the old places and all the birds. He was a drinker, though. He died young.' Lizzie cuts the fruitcake into four pieces and places three of them on plates then gets back up and goes to the window.

Charles takes his watch and sets it down in front of him. 'Normally, I'd have about 15 pairs of lapwings here to show you and I think we've got two. The most we get is 20 but the predation levels are never as bad as they are now and with this shitty weather we've had, we'll be lucky if we salvage any.' As he speaks, he looks over at Lizzie who is hanging out of the window, making clicking noises and tearing pieces of fruitcake for the peacock beneath. 'With the lapwings,' Charles continues, 'I've fledged successfully 500 in 20 years but now they just go out the fence and get bloody murdered. If it's not badgers, it's ravens or town foxes. Forty years ago you hardly saw a raven.' When the peacock has eaten all the cake he starts squawking again. 'You greedy bastard!' Lizzie shouts down at him and he runs away across the grass.

Beyond the window, at the other end of the room, out over the lapwings fields, six mallard are cutting through the warm grey sunset and we sit watching them until they disappear down towards the coast at Aberaeron. 'If you want to see duck at Cefngwyn,' Charles tells me, after he's turned back round, 'let England freeze. You know you get a still night with frost. You've got red scars of sunset and it's moving through to dark.' Lizzie rubs her hands to get rid of the cake crumbs. 'There was one night, Charles, wasn't there?' she says as she perches by the stove, 'you know, when it goes quiet after it's snowed and it's still light but it's dark, that eeriness of quietness.' Charles grunts and nods his head. 'There was no traffic,' she continues, 'and all you could hear was duck on the wing. You know you're looking at stars and the more you look, the more you see. It was like that. There would have been ten thousand duck. Fifteen thousand duck maybe in that sky.'

His nose is smaller and his face is finer but his eyes are the same cold blue. 'That's my uncle,' Charles says, sitting at the bottom of the stairs while pulling on his boots. 'Fleet Air Arms, dead in the North Sea. He was 21 years old.' Above the paintings, two fox heads are fixed to the wall, their fur turning mossy brown. 'My mother had one in 1933 with the foxhounds and my uncle had the other in '36, just before he died.' Their mouths are set wide open, snarling for 80 years. Charles stands and calls down the hall to Lizzie but no reply comes.

Light rain has started to fall, and as we cross the yard a collie runs round and round in its kennel, barking as we pass. In the shed next to it, on a thick bed of sawdust, glowing pink beneath a heat lamp, six fat labrador puppies lie in a circle with their noses pointing inwards. When Charles presses his big face up against the rusted grill and whispers 'babies' under his breath, their tails twitch and, as one, they lift their little heads halfheartedly before settling back to sleep. An old grey radio is fixed to the wall: '*By the end of '85 we had something like 275 AIDS cases and 144 deaths. The health department was taking it seriously and we were trying to communicate that, but there was a lot of prejudice out there.*' Charles listens for a moment before tapping the plastic twice with his finger. 'You see, it doesn't do for me for pups to be lonely.'

The air among the sheds is sweet with the smell of calves, and overhead, bats are on the wing hunting in the half-light. 'With the peewits,' Charles tells me, as we lean over the rail where a heifer is lying in the straw, 'it's their family ethos. A cock might have two wives, fair enough, but he does his share of the duties. They will fight so their young can live. When I see a real gutsy cock taking on a predator, I take my hat off to him.' Charles unlatches the gate and goes to the beast, then bends down slowly next to her. 'She might calf tonight,' he says, as he runs his fingers across her swollen tawny belly then rubs softly at her face. 'I'll come out tomorrow morning and check her at five.' He draws himself upright, stretches his back, then walks into the shadows. It takes me a moment to realise he's pissing in the straw.

When he's finished, he runs his hand once more over the heifer and then we walk back through the sheds. In the yard, the peafowl are going up to roost in an old oak and the rain is starting to come down harder. Charles ducks under a corrugated eave, where a green light is flashing on the wall. As he looks up at it, he tells me about the first book he ever had as a child. 'Historical document now, really. It says in it that lapwings are a very common bird and there's a picture of them nesting between grazing cattle. When the young hatch, the hen is happy there because the grass is short and her chicks don't get swamped. That's the farming job done properly.' Charles strains his eyes to read the number beneath the light. 'That's my pulsator. It takes the electric around the whole exterior. There's a fair hump going through but it should be higher. Something's touching it. Could mean a badger has tried to dig its way in.' Over the years, Charles has found stones that were holding down the bottom of the fence that seem to have been rolled away. 'They're very canny but are they actually deliberately shorting the circuit?' He breathes in deeply after he's said it and shrugs. I'm unsure whether he's self-conscious about how mad it sounds or if he's frustrated that I don't seem to understand. 'What I do know,' he says, as we head in, 'is that when Billy Brock finds lapwings, he won't ever leave them alone.'

Upstairs, in the spare bedroom, a small silver rocking horse casts shadows on the wall, and by the bed a stack of books and a pile of fishing tackle have tumbled together. At the window, silhouetted in the darkness, Charles is looking down over the lapwing fields. He finds it upsetting sometimes being able to see so much. 'There's places across the country where badgers have taken nearly two-thirds of the lapwing nests and I see them coming across, foxes too. It's the same job every night.' Behind the rain clouds, the moon is almost full, and, two fields out, a pond is cast warm grey in the pale light.

First it claws at his arm, then its head starts moving back and forth, frantically trying to take a chunk of flesh with its beak. In Charles's large red hands the crow looks fragile and small. He turns it upside down, then grips its neck with his forefinger and thumb. As he pulls, it flaps twice, then he throws the body into the gorse. 'Here you go, boy.' He reaches down into the trap and reattaches the perch. The decoy bird is looking up at us, its feathers wet with rain and its black eyes forlorn. 'I'm sentimental,' he tells me, running his hand over the bird's back. 'When they've done me a good job, brought in plenty of birds, and they've been with me all summer, I let them go. It's stupid, but it's right.' The previous evening, we stayed up drinking too much whisky. Charles won't go to the local pub anymore, ever since the landlord started saying the Germans should have won the war. 'I'm just in the middle,' he explained, by way of making sense of it. 'My father was a Yorkshireman from Settle and my mother's family have farmed here for a hundred years. I'm not exactly English and not exactly Welsh.' Charles tells me he won't take his cattle to the market at Tregaron either. 'It's not the way it should be. The beasts here should support the local market, but they're all there round the ring holding hands. If you go to bid against them, the animal moves up in price 20 per cent.' My eyes feel tight in their sockets – we went through a bottle between us – and the sweat pools on my forehead then runs down my face with the rain.

We churn away in the mule, Charles raising his voice over the engine as it struggles through the mud. 'There was a pair of crows in one of my main lapwing fields last year,' he shouts, leaning towards me, 'and this one lapwing cock couldn't shift them on his own so I caught them. One of the bastards was getting closer and closer. Another 24 hours, I'd have lost those eggs.' To our left, down in a ditch, three bullocks stare up at us. 'You naughty boys,' Charles calls to them as we stop, 'you know you shouldn't be in there.' Slowly, they turn and wander away, nose to tail, one after the other. When they've disappeared from view, we park the truck under a beech tree and set off on foot. 'You know how many millions of cattle just get pumped full of hundreds of quid's worth of soy from Brazil?' Charles asks. 'Some of them spend all their life in a shed, but it's no life. It's bloody horrible.' As we walk he tells me that as a boy he learned that a cow or sheep can only recognise a certain number of people and to him it's one of the very worst things. He stops and points back and forth from himself to me. 'Do you know me? Do I know you? Are you going to be nice to me? Am I going to try and kill you? Are you going to try and kill me? When one person is working with thousands of beasts, it's just a constant state of psychological flux. That's just fucking horrible.'

Every 20 yards, Charles kicks up a cowpat then stands over it, watching the beetles and spiders flee, while telling me they are exactly what his lapwings want. Ahead of us, at the end of the farm, a further fence, 6 feet high, runs round a pond. 'A fence within a fence,' he says as he disconnects the electric, 'the lapwing's redoubt.' When we've closed the gate behind us, he leans on the top rail and looks back the way we came. 'It's ideal that – short, beautiful grass. They always used that as an ancestral patch and in February there were lapwing cocks all over, and then between everything, without me realising, they started disappearing. It was this fucking goshawk.' He spits the words and then heads on, sticking to the fence, and talking alternately about the pleasure of walking among marsh marigolds and the bird that killed his lapwings. 'It's just a lovely flower. Logic says she should be dead, because there's far more lapwings than goshawks. It's one of my greatest joys, the marsh marigolds. Most counties in Wales do not have lapwings, but they don't give a shit, do they? They don't fucking care.' From the rushes, 30 yards out, on the edge of the pond, a lapwing casts upwards through the rain. 'It's the most horrible day when they stop defending,' Charles says as it cuts low above us, 'because you know then they've lost their young. When you see a crow flying over the field and you don't see lapwings trying to push it off, that's the saddest sight in the world.' Twisting tightly and crying an unearthly song, the bird circles round and round, looking down at us and looking down at its nest hidden somewhere among the grass.

Twenty years ago, Charles wanted to share his birds. He built hides, created a car park, and registered a charity called the Cambrian Lapwing Trust, but he keeps them to himself now. The floorboards have almost rotted through in the last hide that remains, and on the wall beneath a torn poster of a lapwing, '7 MAY' has been written in chalk. 'That always used to be when they hatched,' he tells me, rubbing at it with his finger. 'But I closed it down. Some bastard stole the donation box and then people started letting crows out of traps and I just got fed up.' The wooden slats in front of us have almost seized but he manages eventually to pull two open. 'You see what you can see through there,' he says, and I sit on the last remaining chair. As I look out over the water, Charles tells me about years past when we'd have been surrounded by lapwings. 'It's truly disheartening but we've managed to save a few. There's babies in that patch of dock leaves and I did see two cocks defending the other day.' On the pond, a mother moorhen paddles with her young, a flotilla of fluff fanned out behind her, and further out in the far margins a shoveller calls her ducklings away into the reeds.

Charles leans down next to me, peers through the slat and smiles. 'Bloody well done, girl. Is that seven she's got?' He counts the shoveller chicks twice, getting to five the first time and eight the second. 'The trouble,' he continues, as a gull flies overhead, 'is that the biggest enemies of the lapwing are badgers, ravens and the goshawk, but you're fighting against politicians who want to get elected and their ignorance of the countryside is phenomenal. These people have never had mud between their toes.' The gull drifts across to the far side of the pond and the lapwing cock takes flight. As we watch the bird rising towards the predator, trying to push it off, Charles tells me that all they do is listen to the people who shout the loudest. 'They won't take difficult decisions to achieve a balance.' The gull flies on but the lapwing stays up there, drifting back and forth over the fence. Charles grunts dismissively when I say it seems he's doing all he can. 'I could have killed that goshawk, but I'd have ended up in the back of a police van going to Carmarthen and I'd lose my gun licence. I don't think it's worth that.' Together we push the slats back up and Charles bumps his shoulder against the warped door to force it shut. Just above the frame, a little ball of mud sits under the edge of the roof with dark blue tail feathers sticking out. 'I'd say she's dead on the nest,' he tells me, reaching up and rubbing the back of his hand against it, before waiting a moment, and then lifting it down. Charles holds the swallow in his palm and runs his fingers over its pale breast. 'Don't you think she's beautiful?' he says, as he passes me the bird. 'Five thousand miles to starve in this shitty cold spring.'

But if you listen carefully

*The blackcock that has been hanging around for a few
days came to get its photo taken. It comes to the patio
outside my window and we watch each other as I take its
pictures.*

Tom Pickard, '4th November',
Fiends Fell, 2017

Red wattled eyes and deep purple breast shining, wings outstretched,
brilliant white underside on show and fanned-out tail in the shape
of a lyre. Moth-eaten at third glance, the black grouse will soar
forever, fixed to the ceiling in a glass box. Robert Gladstone stands
beneath it, looking up. He turns towards me and repeats my question before answering, all the time casting his eyes aside. 'When
was it killed? Nineteen hundred, I suppose, or thereabouts.'

We are in the bubblegum bowels of a pink baronial mansion in
the hollowed-out heart of black grouse country. Today, there are just
6,000 pairs left in Britain and populations, in some areas, are disappearing at a rate of up to 40 per cent a year, but several centuries
ago the bird's bubbling song could be heard in every corner of
England. So plentiful were they that along Hadrian's great wall,

running east to west from the Tyne to the Solway, the bones of vast packs of black grouse that kept the legionaries fed have been scratched up out of the ground by archaeologists. At the other end of the country, on Surrey heathland, they hung on until the 1750s and they were still being hunted in Wiltshire at the beginning of the nineteenth century.

The causes of their decline are many. Disturbance at their lekking sites where, in spring, cock birds put on mating displays – rich voices rising and tail feathers proud white carnations – causes significant harm. Changes in agriculture have also been terminal. The demise of the small patchwork farm, where a couple of acres of oats ran alongside beets and hay, has resulted in less food and cover for hiding away from predators. Then, in the 1950s and '60s, land was drained in an effort to improve grazing for sheep, leading to the disappearance of moisture-loving insects, which sustain black grouse in their young months. After the day of the sheep came the night of the non-native tree when moorland was planted with lucrative timber. In the first two or three years, plantations provide ideal scrubby black grouse habitat, but within a decade the trees shoot up, the canopy closes over and the berries, buds and heather growing beneath, which form the bulk of the black grouse's diet, are shaded out and killed.

Silken moustache, thin grey hair, and watery blue eyes that match the colour of the stripes running down the old school tie, Sir Hugh Gladstone's portrait hangs above us. Morning light streams through the greasy dining-room windows and Robert stands on my left, at the head of the table, same nose, same gaze and same sort of age as his grandfather's painted likeness. Fingers clasped to her ribs, Robert's wife, Maggie, lingers in the doorway, looking up at the portrait. 'He was, I suppose,' she says merrily, before turning and disappearing, 'the first Gladstone who didn't really have a job.'

Around him on the walls are paintings of Sir Hugh's forebears, austere renderings of a family who had grown so rich through trade and the toil of the enslaved that Sir Hugh could devote his life to studying birds and shooting them. In 1910, his most famous work, *Birds of Dumfriesshire*, was published, which was followed by a string of lesser-known titles including *Birds and the War* and, in 1927, when he served as Vice-President of the RSPB, *Shooting with Surtees*. As well as his books and regular contributions to the *Scottish Naturalist*, Sir Hugh kept extensive game records, in which he lovingly noted every last detail of every time he stepped out with a gun.

Leather bound, his records are spread across the table in front of us, embossed lettering glinting gold again in the sun. Pale hands planted on the polished oak, Robert looks searchingly at the books before reaching out to pick one from a small pile. He opens it, leafs carefully through the pages and then begins: 'To my everlasting regret, Auchenbrack and much of the ground adjoining passed from under my control in 1911 and since that date, for some . . .' He stops, cocks his head, then places a fingertip on the pulpy paper. 'What is this word?' he summons me. For a moment I fear it may be something pornographic. 'In-scru-ta-ble,' I sound out the syllables and Robert nods. 'For some inscrutable cause, I especially blame the superabundance of pheasants, blackgame have all but totally disappeared in the vicinity.' Even now, over a hundred years after Sir Hugh's observation, there is still progress to be made in understanding the impact of releasing gamebirds in fragile black grouse habitat such as moorland fringes. It is generally accepted, though, that reared partridges and pheasants displace black game and may also be vectors of disease.

At just over 2,000 acres, Capenoch was never a sufficiently large enough estate to satisfy Sir Hugh's appetite for sport, which led to him renting some seemingly peerless shooting on neighbouring ground from the Duke of Buccleuch. Picking up a much bigger book, Robert coughs grandly and turns to the index then murmurs to himself while locating the right page. He reads deliberately as though he's a man dismissing a boy who has asked to marry his daughter. 'The largest bag of blackgame I have seen shot is 114.' I glance across and note the date pencilled in neatly next to the entry: Auchenbrack, 25 October 1910. Robert tells me that it may well have been the greatest number of black grouse ever shot during a single day in Scotland.

Robert wanders round the table and looks out over the garden with an expression of both bewilderment and boredom. I try to fill the silence. 'Do you remember when you last shot a black grouse?' He nods and I nod and then I nod again, and then after some moments he says, 'yes.' I glance down at the fireplace – scrunched paper, kindling and coal waiting for a match. Robert walks past me. 'I need a map.' As soon as his footsteps fade, I reach for the nearest book. Turning through the decades from the 1930s to the '40s, the black grouse column becomes emptier and emptier. Every so often, newspaper clippings have been glued in. Under an entry for January 1945, it notes that Sir Hugh and three companions shot 17 pheasants, 3 woodcock and 10 hares, while a cutting from *The Times* reports that eight officers were killed when a bomb fell on London.

Running my fingers along the top of the book, I feel a raised edge towards the back, and when I lift it onto my lap, it falls open – someone before me has returned again and again to the same page. Glued in there is a sketch of a scene many hundreds of miles beyond Dumfriesshire and Capenoch. A skein of barnacle geese, bunched tight together, fly low over vast snow-swept fen. Down in the bottom left-hand corner, against white frosted reeds, the artist's name is written in blue: 'Peter Scott, 1938.' Years later, in his autobiography, *Eye of the Wind*, published long after he founded what became the

Wildfowl and Wetlands Trust, Scott would admit that most men 'reach a certain stage or age when the old phrase "it's a lovely morning; let's go out and kill something" is no longer funny but obscene.'

Boots sound in the hall and I close the book more forcefully than I'd meant to. Robert reappears and rolls a faded map out across the table. 'We'd drive black game on corn stooks. There were still two or three little farms up there. There were oats and there were meadows that provided nesting habitat.' As he speaks, he moves his outstretched fingers over the inked contours as though remembering the way the birds would fly and then he curls his hand and paces back to the end of the room before turning his head to the hills.

To my left, beyond the window, over the drive and across the Scaur Water, dense forestry cloaks the land. 'Do you think it should be moorland?' Robert turns towards me and I rephrase the question. 'When you look at the trees, do you feel any sense of sadness?' He runs the tip of his tongue across his bottom lip. 'I have absolutely no regrets, because I'm transforming the economics of this estate in terms of the sort of income we have both now and will have.' I glance at the paintings on the wall, great men, portraits side by side. 'Must money always come first?' His expression changes impercep-tibly. 'We should try and achieve a balance, but economics do have to come first. We have a different sort of wildlife now. The deer population has increased a lot.' He catches himself as he says it and then takes a different tack. 'One converts an asset which is producing a very low income through a lot of hard work into something which produces a very large income. We plant the poorest land then we end up with a very large capital sum.'

A clock somewhere in the house chimes 12 noon. It is time for me to go back over the hill. As Robert piles up the books, he looks down at the map still rolled out in front of him. 'It was last year, you know, that I last saw a black grouse,' he says, as he studies it.

'When I was a boy we'd go up the march dyke, perhaps quite ille-gally, and stand the guns there. The beaters would bring the moor forwards,' he points at the faded lines. 'It was a young wood then. There were thorns. I saw three black grouse there last year.' His voice has softened and he stares at me directly. 'I have no idea where they came from. They flew past, right past my head.' Looking up at the portrait, I imagine Robert on the hillside as a boy, standing next to his grandfather, gun loaded and his little cold finger on the trigger. 'What were you doing when you saw the grouse?' I ask, as we descend the stairs back beneath the great stuffed bird. 'Planting trees,' he replies. 'It's all turned over to trees now.'

When I'm finding my coat, there are footsteps in the hall. 'Have you shown him the two-headed duck?' Maggie emerges through a side door. Robert shakes his head and she closes her eyes for a sorrowful moment. 'He was very into freaks but the two-headed duck has been missing for a bit.' To Maggie's left, I notice for the first time a stuffed bear, 4 foot tall, its dark fur the same colour as the stained panelling behind it. Paws outstretched, it holds a tray. 'What are you going to do now?' Robert asks. 'I think I might wander up to see that old lek site where you saw the black grouse,' I reply. I turn away from the glass eyes pushed into the little head to see that Robert is frowning at me. 'You could,' he says, 'but there's really nothing there.'

Halfway up the mossy march dyke, a hare breaks from a clump of bracken and lollops lazily away in the heavy autumn light. The creature knows it is faster than me and it knows nothing of guns. On the hard edge of the spruces it turns to look back and then disappears into the shadow of the trees. Robert was right. Where the black grouse once lekked, nothing remains. I slump down in the grass. It is at its richest, and in a couple of weeks, as summer turns to autumn, it will start to fade. Tilting my head back, I pretend it is spring 40 years ago when, like every spring for as long as anyone could remember, cock birds returned to the very same bit of ground to dance, their voices drifting down the glen on the westerly breeze.

Somewhere across the hill, the sound of a plane cuts over the silence. Then I see it, emerging from the clouds, flying high above the tops. I watch, wondering what the great angular plantations of uniform woodland striding across the land must look like from the cockpit. As it passes out of view, I remember a time when I was a boy and I got lost in the forestry behind Moniaive. For hours, I walked round and round until I found two tumbledown cottages and behind them the remains of a wall. As I followed it, every few hundred yards I came upon piles of stones, gates rotted away to nothing and crumbling pens. It didn't occur to me then that they were forgotten structures where forgotten communities once thrived, people whose lives supported birds that will soon be forgotten.

As the ramp comes crashing down, the crow, pecking at cow shit on the cold concrete, hops up onto a bent gate. The green eyes have seen it 40 times already today: a farmer stood by his trailer, wearily coaxing beasts out into the open for the last push to market. For Patrick Laurie, it started up on the moors several hours earlier, when the sun was rising over the Solway, turning Galloway gold, and the first snipe of dawn were shrieking across the sky.

Pushed together, as far back as they can be, are three black-and-white faces, blinking in the late October light. Patrick taps the metal with his crook and they shudder to a cowering hunch. 'Come on, then.' He speaks to them like babies and they turn their heads towards the familiar voice. But the more he says it, the tighter they push up against the trailer end, until he wades in among the straw to force them out. Each beast must weigh twice what he does and I stand wondering whether I am about to see a man crushed to death, reddish curls trampled by frenzied hooves on the piss-sodden floor. But with stout steps they begin to move and then they canter forwards, buffeted on both sides by cold aluminium fences. A robin has joined the crow and they look on as Patrick shifts from sweetness to anger and back again, pushing the steers towards the stalls.

Midway down, a little man with a big chest in a small shirt, tie tucked into his top pocket, is working away at a red belted bull's tail, hair spray in his left hand and a comb in his right. Regimental, he takes two steps back then halts – a great yellow stream gushes from the beast and flows into the gutter. When the animal is done, the man resumes his position and sets to primping again. Across from the bull there is an open gate and Patrick prods his beasts down the gangway where a boy, all-white coat and steel toe-capped boots, blocks their way, forcing them into the last empty pen.

Elbow to elbow, we lean over the rail, watching breath steam from the steers' wet noses as they look back at us in confusion. 'In the summer,' Patrick tells me, 'in June and July they'll roam right over the hill and I often think about the things that they'll have seen and I wish they could show me. Things I'll never see. They've seen the whole underside of that hill, the eagles and the stags and the black grouse.' He pauses and then answers a question that I haven't asked. 'Four hundred and fifty – I'll not part with them for less than that.' The beasts are starting to settle now, and two of them are sticking their heads through the rails, taking hay from the heifer next door.

To modern farmers who want fast-growing, high-yielding cattle that meet supermarket demands, ancient double-coated 'riggit Galloways' are an unattractive prospect. By the 1980s, save for a few throwbacks, riggits – a local word of Scandinavian origin denoting a characteristic stripe running down the beasts' backs – had become extinct. In recent years, though, a small group has sought to revive them and they are now an integral part of Patrick's struggle to save the black grouse.

Smart green jacket and blue corduroys that swing high above the ankle, a man, grey face shot through with broken blood vessels, stops on his way past and pencils a note into his auction catalogue. Patrick trails off, stealing glances at the scrawl before continuing to explain why native cattle 'are the tool for the job' when restoring black grouse habitat. The trouble, on hills that have escaped the scourge of commercial forestry, is that moorland 'becomes dominated by rank grasses and bracken which overproduce massively in the summer'. The effect is that heather and blaeberry, both important parts of a black grouse's diet, are crowded out. Equally, there is little hope for plants like bog myrtle, which support nutritious inverte-brates. Native cattle like riggits are able to smash up and eat all of that suffocating vegetation, whereas fast-growing commercial beasts are happier, out of the weather in a paddock, huddled around a pile of silage. 'Even their shit is different,' Patrick tells me. 'I don't use wormers and the cowpats are gone within a fortnight. But the effect is that the shit kills vegetation and then gets eaten, allowing new shoots to come through.' Over on his neighbour's land he has seen the chemical-soaked dung of commercial cattle lingering eight months after beasts have been taken off the ground. 'Any insects that try to eat away at it are just killed by the poison.' His riggits are not, Patrick hastens to add, some sort of panacea, but when you factor in the enrichment of the soil provided by them making a mess during wet winters with heavy hooves and the dung beetles pulling their shit down into the earth, the benefits of native beasts on the hill become clear. 'I would almost say,' he continues, rubbing his hand across the rump of the soft-faced beast in front of him, 'that if there are cattle on a good bit of ground in the winter, there'll be black grouse in the spring.'

The walkways are bustling now, men mostly, of all ages, getting a good look before the bell sounds, the ring opens, and the auctioneer takes up his gavel. They nod as they pass. They might not know Patrick personally but they knew his father or his grandfather and there have been Lauries farming in Galloway for as long as anyone can remember.

'I'm troubled by it,' Patrick tells me, drumming his fingers on the gate. 'Troubled by what?' I ask. 'Just to see them thrown in for sale with a load of others. They're mine.' He smiles as he says it and I wonder whether it's the sort of sentiment he would ever confess to the men drifting around us. 'You see this one? I always liked her – this is Wilma's calf.' I put my hand out to rub her face by way of introduction. 'It's funny, isn't it,' I reply. 'People go on about hunting giving you an appreciation of the value of life but it's so impersonal, whereas this, with Wilma's calf, it's all so intimate.' Patrick nods and makes a clicking sound with his cheek as though driving a horse on. 'I had to shoot a cow of mine a fortnight ago. I knew her for five years. I drank half a glass of whisky before going out. She was standing as close as you are, and I shot her between the eyes with a .308. She just stood looking at me because I'd been with her in that field every day.'

Lorne sausage skewered on a fork held in a tight white fist, a man sits in the window of the café staring out at Castle Douglas High Street. 'Don't touch it!' he gets up and shouts at the waitress who answers him without looking up. 'I won't touch it, Tommy.' His lighter has fallen onto the floor and he stoops to retrieve it, clutching the waistband of his grey tracksuit bottoms on the way down. 'Watch my breakfast, Wendy.' An old lady behind the till nods gently, 'Nobody's going to eat your breakfast, Tommy.' Oily black hair and frightened eyes, Tommy pushes past us and lights his cigarette, holding it with every finger of a shaking hand while interrogating the bin men.

We find an empty table and a waitress appears at our side. 'Full Scottish?' Patrick asks me. 'Please,' I reply. 'Two of those,' he tells the girl, 'lorne not link and two mugs of coffee.' Castle Douglas is a town built around its market and in decades past the pubs would have been ringing out, but drink-driving happens less now and the curtains are drawn in the Kings Arms across the road.

'I don't want to sound like a complete wanker,' Patrick begins, 'because I know I can be a wanker about this, but what I put into that stall now is real. It's true blue, bred on grass, fed on grass, and raised out on the hill and still everything else is better and smarter because they've been kept in sheds and given all the feed they want. I haven't got the money to buy a big shed.' A whiskery lady in a rain cape is staring, mouth agape, hooked on wanking. 'Fifty years ago, you'd have put those beasts in and there'd have been a hundred others like them and everyone would have gone "fair enough". I put mine in and nobody even looked. They're small and I wonder now if I'm even going to sell.'

Monochrome, with a thick film forming on the beans, our breakfasts appear. Patrick picks and I try to think of something to say. 'If they're not on the hill, they can't be doing anything for the black grouse.' He shrugs and starts buttering his toast. Diagonally across, beyond the window, a man is opening a small shop, blue painted front and a faded white sign, McCowan and Son's. The face of the man heaving up the shutters is familiar and I suddenly remember that he sold me a fishing rod once, a long time ago. I'd been saving

101

up for two years and he gave me £20 off as well as throwing in a landing net. He wasn't fat then.

Patrick calls for a second cup of coffee. 'I don't want to pull away the whole system of what you've seen there, but I just find it frustrating that because of current rules and regulations you can't treat Galloways as Galloways and do with them what nature intended. Every winter your beasts go back and every spring they come forwards and after about five years they'd be ready to kill.' He glances at my plate – I eat at half the pace he does. 'If you take a five-year-old animal to the abattoir these days, you get heavily penalised. It's almost illegal. Everybody has basically got to get to thirty months.' I cross my knife and fork on my plate and shake my head in ignorance. 'BSE,' he continues, raising his voice, 'it doesn't develop until cattle are older, so the 'Over Thirty Months Rule' came about in '96. It's a precaution that's been a disaster for slow-growing native cattle.'

On the way back up the High Street, the temperature has dropped, and if it was later in the year I'd say it was going to snow. As we walk, Patrick tells me someone could make a grand on his beasts as carcasses if they took them on for a year and fed them up, but they'd need silage, which is something he hasn't got, and he needs the money to pay off the debt on a trailer and for baby things – his boy is eight months old.

Shoulder to shoulder, we cram together, large young men with phones held to their ears and little old men just watching, still wanting to be part of it all. The ring is a sweet stink. Cattle piss and dung mixes with the smell of bacon rolls and coffee steaming in polystyrene cups. Down in the alley the thud of muscle slamming into metal sounds and all eyes turn to face the gate. Two boys in white coats lift the latch and a beast lurches forwards, fear in its black eyes and its mouth agape, a great roar of defiance rolling up from its belly. 'How much here, then? The last bull, grass reared. 3,000 guineas to start, 2,000 . . . 1,500 . . .' The auctioneer ratchets down the price until someone in the room bites. Unseen, a flick of the wrist or a nod and then he starts climbing at a canter: '1,500, 1,600 bid, 1,700, two, 100.' The numbers roll off his silvery tongue until he's exhausted the room and the animal has blown foamy snot all over its face. The beast, head hanging, is led away and the men start to drift off. They've seen what they came for.

The gate opens again and the steers tumble in, looking even smaller after the prize bull. 'Three riggits, the last of the day, how much?' Patrick first rode into Wallet's Mart at four years old on his grandfather's shoulders, and 30 years later he stands at the auctioneer's side, eyes cast round the room for any buyers. '500, do I see 400 bid, 350, 300 bid, three, three, is anyone going to get us going?' Up on the wooden seating a hand finally rises. '310, 320, twenty.' Patrick looks from the auctioneer to the dwindling crowd and back again, 'twenty, thirty, 350.' When a cry of 400 goes up, the panic on his face fades to deflation. 'Twenty, thirty, 450 bid, fifty fifty, fifty, all done here, that's 450 bid to you sir.' The gavel comes down and Patrick sits to watch as his three riggits are pushed out of the ring, stout steps at first and then they canter forwards, cowering together.

At Gretna, a badger lies on the hard shoulder, insides out on the wet tarmac. By Haltwhistle, three crows watch a lamb straying too far from the flock. Four miles east of Durham, down at the bottom of his garden, in a shed by the old line that ran to the ironworks, Tom Pickard sits in his slippers. He was born the year after the war ended and when he walks he uses sticks, but he built his kennel himself, he tells me, just over a year ago, digging right down to the ashes. 'There was this kind of patio-type affair. I was smashing it up. As I got down there was an even more well-built sort of blockage, so I started howking that up and there were somebody's fucking ashes.' He takes his white cloth cap off, throws it down among the books, and starts laughing. 'I was gonna try and get them out but the container was sort of rotted, so I thought, well, I'm fucking leaving that. We've had a lot of bad luck ever since.'

Tom has already had his breakfast, but I left early and he insists on making me some. I resist the porridge but tell him, 'Toast would be perfect. Just butter.' As he gets to the shed door, he reaches into his top pocket and pulls out some photographs. 'You can have these. I took them when I was up there.' In black and white, a tall dyke is tumbling down. The coping stones on top have fallen flat and to stop it collapsing, a twisted length of Rylock fencing has been staked to the ground, a temporary measure, maybe, 15 years before. On one side, a man's footsteps thread through the snow and on the other, where the wind has blown up a big drift, the prints of a crow run in the same direction. In the second picture, in faint colour, a rusted yellow outline of a child's toboggan rises above grey long-lying snow and up behind it sits the highest café in the country, out of focus on the horizon, burnt down since. At the age of 56, his marriage bust and heading for bankruptcy again, Tom took the room above the café and walked away into the most productive period of his life.

In 1964, long-haired, 18 years old, and just married for the first time, Tom knocked on Basil Bunting's front door, saying he'd come because he'd heard Bunting was the greatest living poet. The old Quaker modernist looked at Tom's notebooks and told him to leave the short stories and drama and to focus on the poetry. The meeting was an essential moment in the British Poetry Revival and in time

Allen Ginsberg, the visionary mind of the Beat generation, would call Tom 'one of the most live and true poetic voices in Great Britain'. I only really came, though, to hear about the black grouse and Tom, in his slippers, is standing at the door with the toast.

He sits and I eat. Yesterday, he tells me, looking around at the boxes, he spent some time trying to find the pictures of that lonely bird, but he still hasn't unpacked everything and he had no luck. Tom smiles. 'I think he was just getting blown off course and he'd land outside the window. He was very curious in the way black grouse are and we'd stare at each other forlornly then he'd let me take his picture.' Tom says he knows fuck all about birds and that the person I want to speak to is his old friend Colin Simms, but they are a presence in much of his writing, from the wren 'perched on a hawthorn' singing 'a scalpel song' in his erotic epic, 'Lark and Merlin', to the rooks and ravens riding the wind in the *Fiends Fell Journals*. *Fiends Fell* charts the struggles of bankruptcy above the café and the wonder and brutality of the hills out beyond. He points to my toast. 'You get on and eat your scran, I'll talk.' Part of the reason Tom thinks his time on Hartside Moor was the most prolific of his life was the vulnerability. 'You could walk all day and never see another soul and certainly you could die out there. Just the sheer startling, overwhelming beauty and the terrifying cold in that house. In winter, apart from the black grouse, there were no birds around but I had the companionship of the wind.' Tom looks up then stands and walks to the heater in the corner. He grunts as he twists the dial then shuffles back across the rugs and sits. 'I used to go out towards the end of January every year,' he continues. 'I would hear golden plovers. They tended to nest in the same place and the curlews came at a particular time. Everything returned.' He pauses as I finish my last piece of toast and then adds that 'they were the things that sort of brought the Fells back to life.'

Despite saying he knows 'fuck all' about them, Tom believes his poetry would be 'totally bereft' without birds and knowing fuck all might be the point. 'I hate these words like soul, but I mean something like the moment with the black grouse, it sort of triggers something and I'm not quite sure. It's not knowing what that is, but it gets you into a frame of mind. It enables you to realise something is going on and you want to be attuned to it.'

There is a bit of almost-cold coffee left in the pot on his desk and Tom pours me the last of it then sits back and pushes his hands down into his coat. On the table beyond the window, two coal tits are pecking at some seeds. There are moments in Tom's poetry when he transliterates birdsong, drawing up words and sounds, unknown to him before that encounter. In 'New Year's Day', when he looks out alone over the sun-white hill, 'a growking raven groaks' and in 'Valentyne', written years later when he was living on the Solway at Maryport, a returning curlew cries, 'I'm here, I hear.' Leaning forwards in his chair, Tom says he doesn't really theorise it 'but with birds there's suddenness and a spontaneity. An encounter might suggest a line or a phrase or an image which allows me to compose. It's immense joy and I think birds allow you to meditate on the impossible.'

Since he was a kid, Tom's been scraping through by dealing in old books, and over the years he's amassed thousands, but most of them are to be sold, 'too old for government support but not too old to pay taxes'. He steps across the carpet and leans down, looking along one of the shelves, hands on his thighs. 'Yeah, there was this poet who ran this kind of cabaret. He would be in drag as like Mother Midnight. It was just discovered in the 1950s or something, it's called *Jubilate Agno*.' He picks out an old blue anthology, sits back down, and starts leafing through the pages. 'I'll find it. It's amazing.' After a minute he stops, runs his finger down the page, then begins in a voice twice as loud as his own, 'for the power of some animal is predominant in every language' and then it just goes on:

> Let Gideoni rejoice with the Goldfinch, who is shrill and loud, and
> full withal.
> Let Giddalti rejoice with the Mocking-bird, who takes off the notes
> of the Aviary and reserves his own.
> Let Jogli rejoice with the Linnet, who is distinct and of mild
> delight.
> Let Benjamin bless and rejoice with the Redbird, who is soft and
> soothing.

'And blah blah blah, on it goes. It's just amazing.' He looks up and passes me the book. 'Christopher Smart. You'd like him. He ended up dying in a lunatic asylum.' Tom sits in silence as I turn through the pages and then he tells me about his legs. 'They are totally fucking screwed up and I have to use a couple of sticks if I walk any distance. I'm hoping it will allow me to continue walking on the Fells, but I've got this place now and I can hear the wind outside.'

I'd forgotten about the pie that I'd put down on the floor. 'It's rabbit,' I reply, when he asks, 'bacon and mushroom and rabbit. I made it yesterday. The gravy is a bit thin but it's all right.' He picks it up and smiles. 'That's really good. I was talking about rabbit yesterday. I grew up on rabbit. My mother made rabbit. In fact, I published the recipe. It's in a poem about rabbits in winter. It talks about putting lap in it and stuff like that.' I ask what lap is and he shrugs absently and puts the pie down on his desk, next to a wooden pencil box inlaid with a swallow.

Tom pushes back in his chair and clasps his hands to his mouth in thought. 'I'm not sure you've found much here, but I'll tell you what I realised as well. I was once sitting on top of the Fells behind a dry-stone wall when a ferocious gale was blowing. I was kind of watching this peregrine hunting in the wind and at the same time I was making notes or composing a poem.' He stops and looks out into the garden where a cat is climbing a wall in the sun. He watches and watches until it jumps down before continuing. 'I kind of realised at that point what I was doing, in just attempting to make a piece of art, was comparable with what the peregrine was doing. It was as much a part of nature as what I was looking at.'

As I get up to go, I ask the question I've been meaning to ask all morning. 'Would you mind, Tom, if I use your vignette about the lonely black grouse as an epigraph?' He stands, kicks off his slippers and holds his hands out. 'Use whatever you like and come any time.'

Down in the footwell, rolling back and forth over a woollen baby mitten, an empty bottle of whisky knocks against my boots. It's the first time in three weeks that any rain has fallen and when I climb out of the truck to open the gate, the sweet smell of bog myrtle hangs over the thirsty hill. Half a mile further, up ahead on the track, a small dark creature hops over the earth where the railway line once ran – 1864 to 1964, a hundred years of shepherds and soldiers and lovers heading west to Portpatrick to board the boat to Ireland. Left hand on the steering wheel and the other on the roof, Patrick Laurie leans his head out the window. 'What the fuck is that?' He slows the truck and shouts it again before ducking back in. 'It's jumped up on your side, have a look.' In a crack where wet ferns are growing up out of pink sandstone, the thrush is trembling, its tail feathers twisted. 'I hope he recovers,' Patrick says, leaning across me to look at it, 'but he won't.'

By the time we get to the granite viaduct, pale light is rising in the sky, casting Loch Stroan a hard blue. It was only 10 years ago that Patrick first went out along the line, but back then there were still well-established blackcock leks on the hill. 'There was two of six, one just up ahead of where we are now and then there were several smaller ones, two and threes. We should see a bird today but then there's a fair chance we won't. That's about the par we're on.' On the creamy wet grass above us, the remnants of farms still run across the land and a hay rake rusts red on the side of the track, but the fields have been empty for 40 years. 'It was because of the eagles and nesting hen harriers that it first got officially designated,' Patrick tells me. 'They just pulled all the agriculture off and thought they were making space for nature.' The designation has meant that the land has become 'an island in the middle of an ocean of commercial forestry' but by Patrick's reckoning, 'It's still fucked. You can see just how dense that washed-out white grass is, and the bracken's horrendous. The cattle would just smash that out and graze it down so you'd end up with a much better heather mix, but at the moment, the heather can't come through.' Patrick thinks it's just been quietly forgotten that it was ever a site of special scientific interest. 'Scottish Natural Heritage went for a strange 1970s-era technique of just doing nothing. It was quite well known once, but all the birds anyone came up here to see have declined and most people don't even know where it is anymore.'

Patrick stops the truck with the handbrake and lets the door swing open behind him as he gets out. In the past few years, he tells me, as he sits on the bonnet, the landowner has begun to realise what a state the place is in and has started to think about getting some farming back on. With the 40-year agreement coming to an end, Patrick's been allowed to take the grazing across 200 acres but he doesn't think it's enough. 'Currently I'm only allowed to graze ten cows, which everybody agrees isn't even touching the sides. It's laughable. They're worried that the cattle with their heavy feet will poach the vegetation. It's written into my contract that if there's any kind of damage, my lease is terminated.' He lifts his glasses into his thick curly hair and scans the hill with his binoculars. 'But surely,' I ask, 'poaching the ground is exactly what the black grouse need?' Patrick raises his eyebrows and carries on looking up ahead of us at the horizon. 'Yeah,' he replies eventually, 'and the curlew and the lapwing.'

Fixed with four steel cables, a radio mast 500 yards up the hill sticks out of the ground. 'People used to say they're predictable,' Patrick shrugs, 'and leks did last for generations, but when they get down to really low densities they can move about a lot.' Three mornings previously, just to the left of the mast, he saw a lone blackcock calling at dawn, but beyond the chiming willow warblers the hill is quiet. 'What I'm looking for,' he says, 'is that little fleck of white tail feathers. It's like a blink in the sun.' He waves his fingers in an imitation of the ruffling white carnation, then stops and turns his ear to the hill. 'At this kind of distance, you're really just listening for the pitch change after that bubbling. You can't always hear a note, but if you listen carefully you can hear the pitch change on the air.' He stands, kicks a pebble by his feet into the wet verge and makes a rasping sound. 'Occasionally you hear that too, like a harsh crackle.' He sits for a moment, then smiles and makes the noise again, like the static in a storm rising from his throat. 'When I was young, I always thought it was stupid that blackcock would go back to the same lek even if there's nobody there, but the reality is they might display somewhere for an hour, and if it's not working they'll fly 2 or 3 miles, maybe display for ten minutes, maybe then fly another mile. They're trying their best.' Behind us, above the loch, a deep vein of red is running across the sky and two greylag geese have taken flight over the still water. 'Part of it,' Patrick continues, 'is just how discordant they are. They're just astonishingly smart. At this time of year especially, in this very kind of washed-out environment, you've got this absolutely glossy flash, that glossy white, they're extraordinarily gaudy.' He turns his head to the hill again, listening, and then tells me we'll drive on. 'The next lek is like a mile over, so he might have gone to see the other ones. You've got to have that interconnectedness. Nothing drives the black grouse population down like fragmentation, and nothing fragments a habitat like commercial forestry.'

For a minute, as we grind along the track, the land is cast in a dull golden haze and all the edges blur into one and the geese become part of the sky. Then I realise my glasses are still on the roof and I can't see. 'This type of ground hardly exists anymore,' Patrick says, when I've fetched them down and we get going again. 'The people who've got it don't mind you grazing it, but you know they're always looking over your shoulder at opportunities to make stacks of money. One place I've got cattle, they're about to take a small plantation out and the owner will be in for £400,000 when it goes to the mill.' Up ahead, just in front of the charred remains of the railway fence, three Scots pines are growing at the edge of the track and beneath them, covered in moss, a concrete platform is all that remains of Skerrow station. There are old photos of hill folk waiting for the train but it was mainly a place where the steamers stopped to have their tanks pumped with peaty water from the loch below. The old stationmaster lives down in Dalbeattie now and his cottage, where black grouse lekked in the garden, is gone. In the 1980s, the Ministry of Defence was looking for a vast swathe of nothing with something the army could blow up, for practice, and it was thought Skerrow would do.

'It's all part of it,' Patrick says, stopping the truck and winding his window down, 'the loss of the people and the cattle culture.' He pauses for a moment when a cuckoo calls, then looks at me. 'I have a clear understanding of what I think Galloway is and it's just undergone these huge changes in the past 50 years. In lots of ways it's intangible, but you can measure that change by the loss of black grouse. There's like this huge, massive shift.' The cuckoo starts up again and Patrick raises his fingers to his mouth, running them across his lips. 'Black grouse are just these talismanic birds,' he says slowly. 'I place everything on them. My family has been intimately involved with them for generations.' He leans across me to wind my window down and tells me he can't imagine what it's like not to feel as though you're from somewhere, that you're truly from a place. And then the car alarm goes off. 'You fucking arsehole!' he shouts, banging his left hand on the dashboard and pulling the key out. 'I just end up sounding like a fucking idiot and like I'm reading too much into it. Maybe it's working in isolation. So many people will say it's just a bird but it's not. It's not.' He gets out, walks to the other side of the track and stands in the moss, steam rising in the morning light beyond his boots as his piss hits the rushes. 'I have a feeling there's something there,' he says when he returns. 'There's something over that cuckoo.'

At the end of the loch, beneath a steep bank, Patrick stops the truck and reaches into the back for his flask of coffee. 'What we'll do is, we'll just climb quietly up here and put our heads over the edge. If we look back, it's more or less where I thought I could hear that blackcock.' Down on his haunches, he wades slowly through the bracken, then stops with his elbows on an old stump. Glasses in his hair, he peers over the ridge with his binoculars. For a moment he scans the hill then sits, with his knees pulled up to his chest, and pours himself some black coffee. 'I spent 10 years in a state of almost desperate upset, and I think what I'm trying to do now is come to terms with the reality that they aren't coming back. There's nothing we can do fast enough.' A snipe somewhere overhead, unseen, is drumming and Patrick turns his head towards the sound as he whispers, 'What's maddening is that if you've got black grouse on a place you need to legally protect them, but if you don't, you don't have to. I guess they think, why would you protect a species that isn't there, but their whole existence depends on their ability to move into places they haven't been before.' He rolls the coffee towards me along the mossy ground. 'Do you want some of that? It'll be cold.' The flask is baby sized, with lambs in pink, yellow and blue running round it as though they're chasing each other. Patrick was wrong, the coffee is hot. 'So every time you lose black grouse from somewhere,' he continues, 'the landowner is like great, we'll turn it into forestry and it's just shutting the door behind them every time they leave a room.'

One afternoon, a couple of years ago, when he was passing through Dalbeattie, Patrick stopped in to see the stationmaster. The old boy didn't invite him in, but standing on his doorstep, he told him that on summer days he used to cycle down the line to check for fires started by coal from the steamers. Every couple of hundred yards he'd be picking up black grouse that had been hit by passing trains while gathering the grit they need to grind up the heather in their crops. There were so many then that they were just another part of the scene beyond the train window. Patrick shakes the drips from his mug into the grass, then lifts his binoculars to the horizon. 'Fuck's sake,' he says, breaking out of a whisper, 'there's a fox up there, a big dog fox coming down over the bracken. He's probably the thick end of half a mile away.' Patrick passes me his binoculars and I ask him how much of a problem the foxes are, as I watch it padding down the rocks, stopping every 10 yards to sniff the wet air. 'I don't know,' he replies, 'I've kind of been reappraising my understanding of predators. I think foxes are a major problem, but at the same time there are people who just kill them recreationally. It's kind of macho bullshit.' The fox disappears out of sight among the grass. 'Is it okay to just imagine we have a limitless supply of foxes? At least fucking think about why you're doing it. I'm being a grumpy prick, but I'm sick of trying to justify a lot of really shit predator control and of people piggy-backing off what is really effective predator control because they just want to shoot foxes.'

Somewhere behind us in the trees, a roebuck barks three times and I turn to look back over the loch towards it. The pink-seamed clouds are darkening and rain is starting to fall on the water. Patrick pulls up a clump of grass and puts it in my hand. 'If you look at a greyhen, a female, a lot of her colours are based on some of these colours. It's supposed to be part of their habitat but not a vast block of it. The cattle would basically pick the guts out of it. They'd eat the grass seedlings, which knocks it back and means it doesn't smother the heather.' Lifting off a burn below us, a pair of greylags circle noisily over the hill. We stop talking when they pass and then sit quietly for a while after they've gone.

A five-point turn and then back down the line with the rain coming down hard on the truck's canopy. At the Castle Douglas roundabout, Patrick apologises. 'But then even if we had seen one,' he adds, 'all you're seeing is this confused, disappointed bird that will never fulfil its original function. I wonder what that's like sometimes to be born with all these inherent behaviours, to want to work constantly as a social animal, but it's only you and you're broken.' Sharp right on the old road to Dumfries, across the river, and up the track to his white cottage. 'Will you have an egg for breakfast?' he asks, as he walks across the yard. 'Please,' I reply. 'Fucking move!' he shouts at the bantams as he ducks into the shed at the edge of the yard.

A spider, on the whitewashed kitchen wall, runs up and then falls back down again behind the sofa above where Tina Laurie sits, left hand twiddling her brown hair and a small boy held with the right. 'Any luck?' she asks. 'No,' Patrick replies, 'all the way out beyond Skerrow and we didn't see a thing.' The boy looks at me curiously with his little blue eyes and holds his hands out to his father. Patrick picks him up and he nuzzles his auburn hair against the pale neck. 'We've got bacon from a pig that became violent,' Patrick says, as he stands over the range. 'How violent?' I ask. 'Worryingly so,' he replies. 'We couldn't let Fin out in the yard. Some mornings I didn't fancy it.' One arm round the toddler and the other over the pan, Patrick cracks the eggs, and fat and smoke drift across the kitchen. When he goes to cut the bacon he puts Fin down, who then stands unsteadily for a moment, staring at me. 'When he's older, do you think there'll be any black grouse left in Galloway?' Patrick shakes his head. 'No, and nor in most other places the way it's going.' The boy totters round to his nursery-school lunchbox and grabs his biscuits before wandering across to the old black Labrador in the corner. Smiling and clasping the digestives between his little fingers, he holds his arm out in front of the greying muzzle. The animal eats them in one, before licking the small palm with its big pink tongue. The boy laughs and laughs and then the dog stands and in doing so knocks him to the ground. Tears rising in his eyes, he cries in silence.

The souls of our dead

With blistered heels and bones that ache,
Marching through pitchy ways and blind,
The miry track is hard to make;
Yet, ever hovering in my mind,
Above red crags a kittiwake
Hangs motionless against the wind.

Wilfrid Wilson Gibson, 'The Kittiwake', *Battle*, 1915,
a boy at war thinks of home.

Eighty-eight years later the activist, Peter Tatchell, would write that Captain Fitzgerald and Lord Kitchener died in each other's arms. Of the twelve sailors who survived when HMS *Hampshire* went down off Marwick Head in 1916, nobody saw anything, but I like to think it's true and that the man on the posters drowned in love. For days afterwards, bloated bodies in cold wet uniforms swept in along Orkney's western shore. Fitzgerald's corpse was sent south to be buried in Surrey, but Kitchener's was one of the 625 never to be found.

Two hundred and eighty-five feet beneath me, the bright blue sea breaks white against the cliffs. The rock is hard and cold on my knees but I can't bring myself to stand so close to the edge. I look round, up the path, back the way I came. The same sandy-coloured rabbit is still standing on its haunches, ears up, watching me. When I turn my head to listen, I can just hear them, the squawk of kittiwake chicks rising over the roar of the water, crying out to be fed. The cliffs would have sounded different when Kitchener's men were dying. In 1891, when *A Vertebrate Fauna of the Orkney Islands* was compiled, the authors wrote that it would be 'a waste of time' to go into much detail on kittiwakes, as they bred prolifically wherever there was any suitable habitat. A mile out, where the sea turns from blue to grey, a fishing boat bobs and on deck a figure leans over the side as though he is dropping a creel.

At Marwick Head, the kittiwake population is now less than 10 per cent of what it once was, while up in Shetland and out west on St Kilda, cliffs that wailed with the little white gulls have fallen silent. There are still well over 300,000 pairs that breed around the British coast every summer, but the rate at which we're losing them, largely as a result of the disappearance of the sand eels they feed on, confirms that the seas are changing.

I check the time. It is 8.43, two minutes before the mine was hit and the lights went out. With a hole torn through its hull, it took HMS *Hampshire* just fifteen minutes to sink beneath the waves.

Ingrid Jolly, small pale hand holding her walking stick aloft, stands halfway up the garden, smiling in the sun. 'You're very welcome. Everybody is very welcome but just say it in your head.' She pauses and glances behind her, as the door of the house opens, then turns back and repeats it slowly. 'In Orkney. It's in Orkney. Not on Orkney. Never on Orkney. Just do it in your head. Just say it until it sounds right.' Her hair is silver and thick, her old eyes bright blue, and as

I lean back on the bench beneath the beech tree, looking out over Kirkwall while mouthing the words, she starts laughing like a child. 'You see. You hear it? You're in Orkney. We would never say on Orkney. Everybody is very welcome here. Unless they maybe think they're going to come and change the world.'

An old t-shirt, with a faded picture of a crab across it, tucked into beige trousers that swing a couple of inches above the ankle, Billy Jolly appears at the door. He stands for a moment, blinking in the light like a salty mole, then stumbles down the stone steps, holding his Zimmer out in front of him, as though he's driving a trap down a steep slope and the pony's about to bolt. Ingrid points her finger at my chest and shouts over to him, 'Patrick, he's come from Dumfries to ask you about the sand eels and the kittiwakes and the fishing.' She says it as though it's one of the funnier things she's ever heard. Shuffling his feet and shaking his head, Billy comes slowly across the grass. When he gets to where I'm standing he looks up at me, peering for a second or so into my face, then hunches over on the Zimmer. 'The sand eels and the kittiwakes and the fishing.' Twice, he slowly repeats Ingrid's words, as if he's asking himself a question, then grunts thoughtfully, as though he's found the right answer. 'The thing is they become too good at it, the fishermen. Shot themselves in the foot. It was a good situation until they realised they could get a purse net over and catch a hundred times as many as what they could with all the fishing from small boats, so that they killed, they killed the whole thing completely.' Billy stops, looks up, and gestures down towards the sea, then clasps his hands back on the grey handles. 'That was the industrial fishing. We used to get a lot of the Danish boats. They would only keep the very best. All the rest, tonnes of sand eels among it, down into the hold for fish meal. Bloody terrible some of the fish that went down, absolutely crazy. Then dried off, then in a powder, then fed to swine.'

When storms rolled in and Billy was a younger man, he would head down to Kirkwall harbour to wait for the Danish boats to appear. Anything they had, which hadn't yet gone into the hold, he would buy cheap and send down to market in Aberdeen. 'I've seen maybe three of my boys and myself down there,' he tells me. 'It was always hellish weather, sleet and wind and cold and wet, and we'd be doing deals with them.' He looks up brightly, smiling with crooked teeth. 'Soaking wet. We'd be soaking wet, absolutely soaking. We weren't caring because there was money involved, you see.' Billy suggests going to his garage and I walk slowly behind him, small steps down the garden path. Where the ground drops away at the end of the lawn, he stops, and his Zimmer falls out from beneath him, knocking over a watering can. He grabs at the wall and turns slowly to face me. 'I'm sorry, it's just the only way,' he says, looking at his feet. 'It's easier sometimes to go backwards than it is to go forwards.' Down the short flight of stairs, Billy steps one foot trailing after the other. From among the raised beds, Ingrid shouts down that the dishwasher isn't working either. 'Sometimes I think washing dishes by hand can be just as good,' I reply, and she holds her hand to her forehead, laughing. 'No, I'm the dishwasher,' she calls back. Billy, standing on the bottom step, shakes his head and smiles.

At one end of the garage, a great mound of wood is piled up almost to the ceiling, and at the other, in front of us, small stools and tables are stacked up neatly one on top of the other. Billy counts them for a moment, under his breath, then tells me, apologetically, that he's in pretty short supply. 'I spend a lot of my time in here if the weather's not great, getting the stools made, but it takes me a long while to do jobs that I used to be able to do just in no time at all.' He pushes his Zimmer across to his work bench and perches on the edge.

When Billy was 14 years old, his father sent him to Lossiemouth to learn to fillet fish. Up until that point he remembers thinking that his dad, who worked as a butcher before becoming a fishmonger, knew what he was doing, but he soon realised he had little idea at all. 'He used a knife, as a butcher you see, but to be quite honest, he didn't realise what a rubbish filleter he was until I came back and showed him what he was doing wrong.' When I tell Billy I once went to Billingsgate Fish Market, he groans in wonder and nods. 'I never been. Oh, I would have loved to have been. See, I always wanted to go down but I just never got around to organising it because I was quite good at filleting and they had what they called "The Young Tradesman of the Year" or something and it was competitions.' He pauses, mid-story, reaches for his Zimmer and drags his feet over to the stools. When he gets to them, he stops and turns to look at me. After a moment, he lifts a small varnished one down off the top and places it on the dusty concrete. With his dry red hand he pats the top of it twice then goes back to the work bench. 'Boys would be coming from all over Britain and I was desperate to go down. I could fillet cod, ling, tusk, I could truss a chicken. I could do all the things. I could see the list every year.' He holds his finger out in front of him and runs it down an imagined sheet of paper. 'I could see the list and I would think I can do all that no bother at all, but I just never got round to it. Oh, it would be a really interesting thing.'

Somewhere, up above us, a great metallic roar burns across the sky. Billy puts his hands to his ears as the jet grows louder and louder, thundering to a crescendo over our heads, before fading out across the sea, leaving the quiet afternoon and the starling song. 'But with your sand eels,' Billy begins, putting his hands back on the Zimmer, 'I wonder if it's also something to do with temperature. They just got less and less and less you know.' He starts telling me about an island he used to go to, then stops and closes his eyes.

'That's terrible. Och, sake, I forgotten the name of it.' He tries a few different names but shakes his head after each of them then starts laughing and points at his forehead. 'When you get old, this up here doesn't work the same as when you're younger.' A couple of moments later, it comes to him and he says it three times. 'Burray, there was Burray and there was Green Holm. I would go there ringing the kittiwakes and the terns. I had a small boat.' Whenever Billy set off, the boys in the shop would ask if he was away to ring their necks but he tells me he enjoyed it and he went regularly, 'until the birds got less and less'. He leans heavily on the Zimmer with his left hand and holds his right hand out in front of him. 'The sand eel were a very valuable source of nutrition,' he says, as he rubs his fingers together, 'soft and oily, very good fish for a small bird.' Billy can't recall just when it was but he remembers that sometimes 'small fish with a hard scaly surface, very like the sand eel in shape', would appear. 'These cursed things choked the young birds you see, completely killed them and of course the adult kittiwakes didn't realise this and they were stuffing these things into them thinking they were feeding them with sand eels but they were killing them.'

The stool is child-sized and Billy sits above me looking down solemnly. 'That's terrible forgetting the name of that island,' he says, 'that really was a part of my life.' For all that his father was hopeless at filleting fish, Billy tells me he was always a far superior musician. 'He was a very good singer, a tenor singer. He played the piano. He could sight read. Played the bagpipes. He played in the Kirkwall City Pipe Band.' Billy stops to draw breath then continues. 'He played in a dance band. He could harmonise just perfectly. He played the saxophone. Yes, yes, he played the saxophone very well and he was what they called the crooner.' He looks down at me, seemingly wondering if I know what a crooner is, before miming a snippet of song. 'All I do,' he shrugs, 'is play the mouth organ and I sing a bit.' Back in the 1970s and '80s, Billy and Ingrid regularly performed pieces by Ingrid's cousin, Allie Windwick, a poet and songwriter who worked as a stereotype operator on *The Orcadian*. 'Dead a long while ago now,' Billy tells me, 'but they were all songs about things that happened in Kirkwall, situations that there were. He was very clever. There were things like 'Butter on the Bow' and 'The Peedie Pakistani'. Billy chews at his lip in thought then says 'that sounds very racist but it's exactly not. If you think of the old days that tinkers used to come around, you know selling stuff. I mind them fine coming to Kirkwall and especially out to the farms. They'd just open their bags and there'd be anything at all and so they eventually dried up but the Pakistanis suddenly appeared.'

Using his Zimmer frame Billy pushes himself further back onto the work bench. 'Eventually, the Pakistanis went as well but they were very nice guys, the tinkers, and I still know some tinkers here but they are just settled now, you see.' Two or three years ago, he can't remember how long exactly, Billy was driving in from the other end of the island when he ran into one of them walking along the road. 'I pulled over and I said to him, do you want a run into the town? He said aye that'll be fine, so he got in and we started yarning and he said I've just learned to read.' Billy shakes his head. 'He's retired, this man. He cleaned dishes all his life in the Kirkwall Hotel. He was a very good worker. I said to the guy, you've just learned how to read? He says aye. Then he says to me, I'm reading books and he says he never realised what a fantastic situation it was.' Billy sucks the air between his teeth and stares down at me. 'He never got around to learning and now he can and he's got this great gift, this fantastic situation.'

We sit in silence for a while and Billy closes his eyes, murmuring under his breath. I'm about to suggest I should go when he says,

'There's a poem Allie wrote, describing himself, he's a terribly modest man. He describes writing, the work of writing being similar to a farmer building a byre. "Me Dydoes!" it was called. If I get the start of it, I'll get there no bother at all.' Billy looks up at the corrugated roof, and holds his hand to his mouth, then eventually tells me he's got it. He shuffles off the bench, stands tall, holds himself up on his Zimmer and begins.

> *My music's like the dairy byre*
> *The fermer built tae had his kye:*
> *Wi' concrete block and fower-by-two*
> *And many a pund o' nails forby.*

> *He biggit early; laboured late;*
> *Times doon below, whiles up abeun;*
> *And grudged a meenit aff for maet*
> *So grett his urge tae hae it deun.*
> *At last the sneck wis on the door*
> *And milkers in, each tae her staa'*
> *When neebors clap't his back and said:*
> *'Feth, boy, bit sheu's no' bad at aa'!'*

> *Alas! There cam' an uncan man*
> *Wha h'mm'd an' heyed an' spok' a haep*
> *On Toon and Country Planning bruck,*
> *And measured wi' a muckle taep.*

> *'No! No! This building will not do!*
> *One needs a permit now, you know!*
> *The door's not tight, the floor's not right;*
> *And look – you've built it far too low!'*

Boy! Wur me dydoes, irr wae no' —
Me and me song, him and his byre!
Wae baith began without a plan
And feth, wae should hiv biggit higher!

Billy stacks my stool back with the others and I follow him out of
the garage into the early evening sun. At the bottom of the steps he
stops and opens the door to the greenhouse. Just beyond the door
there is an old enamel basin held up by bricks. He bends down,
puts his hand to the water and three fish rise to the surface. 'I was
busy working in the greenhouse one day,' he tells me, 'and I thought
what a boring thing this tank is so I went straight out and I bought
myself four fish.' With his finger he traces the largest of them as it
swims round and round. 'Did you ever know a butcher called Porky
Horne, who wrote poetry?' I ask. Billy looks up. 'Yes, David Horne.
He had a shop in Kirkwall. He wrote a lot of poems. One about
the Ba. I used to have it in my head but it's gone now completely.'
Billy turns back to the fish but they've disappeared down into the
grey water. The fourth one died not long ago and Billy buried the
body beneath the tomato plants.

In spite of the summer sun, a wild wind rushes up Albert Street like
a wounded bull, and across the cobbles from the butcher's shop the
ivy chimes with starling song. The sign above the door, bolted onto
the grey harling, reads 'Donaldson's', and all that's left of David
Horne are six sausages, wrapped in clingfilm and polystyrene, in the
back of the shop between the cans of juice and the pink sticky ribs.
David died in 1940 and some 40 years after that, the Donaldsons
bought the business but have been commemorating the poet-butcher
with their 'Porky Horne' sausages ever since. The owl-like lady at the

counter doesn't look up from the till when she answers. 'We don't do the Porkies as a sausage roll, just what's there.' I buy a peppered steak slice and a cut-price Irn-Bru, 50 pence on account of being past its sell-by date, then I walk up the street towards St Magnus.

Scaffolding is fixed across the pink sandstone front of Britain's most northerly cathedral and somewhere in among the metal and wooden boards, a heavy drill whines away. Two legs, in orange reflective trousers, hang down from a hatch but the rest of the body is out of sight. On the wall, set back from the pavement, I sit and eat. The pastry is thick and hard, and every couple of bites I run my tongue round the inside of my gums, ploughing up fat. Twenty yards down from me, a young woman, on her lunch break, draws her tongue along a cigarette paper. We look at each other, then we look away. A herring gull, seemingly sensing my disinterest in the steak slice, swoops down onto the pavement in front of me and holds its wings open for a moment like a cormorant. I take two more bites and then, when I'm sure the girl isn't watching, I drop the rest by my feet and wander up into the churchyard. For a while I watch the rooks, drifting above the sycamore trees in dark swirling clouds, and then I wander through the graves, but there are no Hornes and the door to the cathedral is locked.

Small and pink, the little boy's tongue sticks out of the corner of his mouth. With one hand he holds his colouring book down on the table and with the other he works an orange felt tip back and forth, carefully staying inside the dark outline of the tyrannosaurus. Next to him, a couple who I assume are his parents, sit in silence. Her eyes are swollen with tears and he looks away, as though to say there's nothing else he could have done. At the other end of the café an old man, out for a baked potato with his carer, is holding his false teeth up to the light. He looks at them, groans loudly, then sings them a happy gummy tune. 'Stop that!' the carer says, as though he's a naughty child, and the old man brings the dentures down hard on the table next to his plate.

I have to ring twice before I get an answer and when I do I'm not very sure Dewar knows who I am at all. 'It's Patrick, Debbie's son, as in Judy's grandson, as in Judy your sister.' He says 'Debbie's son' twice as though it makes no sense, because I'm surely still six years old, just the same as I was when I saw him last, 23 years ago, at his 70th birthday. 'You're in Orkney.' He says it twice and grunts with approval. 'Very prosperous. Prosperous island indeed.' I tell him I'm up to see someone about kittiwakes, but I thought I'd maybe go and see where his family used to live before they went south to Glasgow and started manufacturing hot water valves. At one point, religious mania swept through the workforce and the unrest caused the business to tumble into a crisis for a while, but at that party, over two decades ago, Dewar was back in his big car and the grown-ups all said everything was well again.

'I went to the butchers,' I tell him down the phone, 'Porky Horne's – I had a steak slice.' He grunts like I've just chipped my ball into the rough and says he's sorry but those aren't the right Hornes at all. 'Not related to us, not as far as I know. Twaat in Birsay is where you need to go. My grandfather was a schoolmaster and preacher there. You wouldn't know but there was something called the Great Disruption in the Church of Scotland, it was one of those periods of religious mania, a crisis. That's when my family left Orkney and went south.' Outside, thick clouds have gathered over St Magnus and raindrops run down the greasy café windows. When the lady passes, I hold my coffee cup up and ask for another.

127

Thin mesh curtains hang behind the windows, the big door is bolted shut, and above an empty flowerpot a trace of white lettering, on a rotting piece of chipboard, reads 'Twaat Church, Birsay'.

The old schoolhouse, newly painted, sits across the road from the kirk. When I wander by, a man is asleep in an armchair with the telly playing away in front of him. I stop for a moment and look in, wondering whether at some point my great great grandfather stood at one end with a piece of chalk in his hand while lots of mucky little faces stared up at him.

The sky is still dark, but out in the west a bright pink haze cuts over the sea, and above me as I walk along the lane, swallows are on the wing hunting insects in the heavy air. On the verge, every couple of hundred yards, there is a low squat cottage, home to a family maybe when George Horne was in Twaat but boarded up or derelict now. Behind the kirk, a sandstone wall runs along the field and I climb over and sit with my feet in the nettles. In the distance, a tractor drives slowly back and forth across a steep field with a man, hunched over, following on behind.

The view down across the loch and out over the sea is one my great great grandfather must have seen thousands of times, and the swallows above me must all be descended from birds that flew above him once, on high summer days when he walked the same lanes. For half an hour I sit, wondering whether this is the bit when I'm meant to feel some sort of wholeness or like I've discovered a part of myself. At half ten the geese take flight, the slate clouds fade and the sky starts to lighten. It won't be dark until after midnight and even then, there will only be a few hours of half-lit stillness before dawn rises.

On my way up to the cliffs, cutting through sandy fields beneath a cold, smudgy sky, I tread too close to a curlew nest and the parents take flight. In the breeze above me, they cry out, a shrill insistent song, then they drop down over my head, criss-crossing back and forth to drive me away.

Up at the top, beneath the squat grey castellated tower that memorialises the drowned field marshal and his men, a woman shelters from the wind, a hat pulled halfway down over her face and her knees tucked up to her chest. There isn't anybody else around, so assuming it's Ilka, I climb up on the style that crosses the fence and wave. She gets up slowly and with one hand holding a tripod over her shoulder and the other cupped to her mouth, she wanders down the hill towards me, pale and cold and smiling. 'You are Patrick.' She says it as though I was possibly uncertain, then tells me, laughing, that she thought she'd get out at 6 a.m. to do a food survey on the guillemots but it was 'far more chilly than expected'. We carry on along the clifftop, heading away from the memorial, and Ilka explains that one of the mysteries of the kittiwake, compared to a number of other seabirds, is that it's much harder to monitor the quality of sand eels they're feeding to their chicks. 'You can't really learn so much about that,' she says, lifting her hand to her mouth and making a spewing motion, 'because kittiwakes regurgitate food, whereas with the guillemot, for example, you're measuring in proportion of how much of the sand eel is hanging out of the beak.'

Some honey-coloured and some black rabbits run ahead of us through the purple sea thrift, and all around their burrows, stripped bare by the bonxies and bleached in the sun, the bones of those that didn't run fast enough are strewn across the ground. As we wander over the warrens, Ilka tells me to be careful. 'Patrick, this is very fragile.' I look down to see a little pair of eyes staring up at us from the darkness, waiting till we pass. In 1986, a monitoring programme was established to collect data on the 25 breeding species of seabird in Britain and Ireland. Every three years, in an attempt to understand how they're coping, counts are taken at 10,000 sites. As well as spending months in the field surveying, Ilka produces the final report and is responsible for ensuring that consistent standards of data collection are maintained throughout.

Cutting left, over a sunken lichen-covered slab, Ilka goes on ahead of me and sets up her spotting scope on a rock that juts out where the cliff falls away to the sea. 'It's actually really simple,' she shouts over the wind. 'At the end of May, when the kittiwakes are returning, I come and do just counting on the cliffs to understand first the population size changing.' For a moment she looks down at the waves, then turns back and smiles as though waiting for me to join her. Feeling more and more unbalanced, I walk steadily out across the rocks, then I sit at the edge, where the red stone is being worn away by rain and time. A hundred yards ahead, on the other side of the inlet, two rows of fulmars are spread along the cliff, and beneath them, smart white bellies and thick beaks blunt and black, razorbills are standing together looking upwards as one. Further down, their beaks to the wall and their backs turned towards us, guillemots are gathered, guarding their young, and further still, just 5 yards or so above where the steely sea rolls in and breaks, kittiwakes are hunkered down, their chicks warm against the wind. 'It's fascinating,' Ilka says, while adjusting the spotting scope. 'It does depend a bit on the cliff site, but this order here is very typical.'

This year, so far, the kittiwakes have been okay, but as the world warms, seas are getting stormier and three years ago, when Ilka was last in Orkney, they were all washed out. 'The lower section,' she tells me, pointing down at them, 'before they could even hatch, was completely washed away.' Ilka sits next to me, reaches for her rucksack, and pulls out a large black-and-white photograph. 'This is how I go about it,' she says, passing it to me. 'This was taken in the late 1980s. It has each of the different nests and I've marked them all.' Every year, after wintering out in the ocean, kittiwakes will usually return to breed at exactly the same bit of cliff they were born on. 'I count five times over three weeks,' Ilka tells me, 'starting this year from 26 May, and from that I get an average.' Beneath us, returning from sea, an adult kittiwake cuts in through the wind and the bottom of the cliff bursts into life. Ilka stops and turns her head, lost for a moment in the sweet screech of hungry chicks.

After putting the photograph away and fastening the ear flaps on her hat up over her head, Ilka stands and goes back to the scope. 'They are almost all on young at the moment,' she tells me, panning along the cliff face, 'so it's really quite lovely. My first chick this year was 10 June but I have a feeling that everything was a bit late because of this cold.' Particularly when the weather sets things back, young birds that are returning for the first time build what are known as trace nests. 'They just carry on working,' Ilka says, looking up from the scope. 'They get a lot of material but they never end up sitting on the nest.' Ilka laughs and laughs when I ask if it's because they want to build them perfectly but they don't quite manage. 'This is not how kittiwakes think,' she eventually replies. 'It is believed to relate to their condition when they arrive.' After she has counted the number of kittiwakes that have returned and has established a rough idea of how many pairs there are, Ilka moves on to productivity monitoring. 'What I do now,' she tells me, 'is try to establish the output of what two adults produce throughout the breeding season. At the moment, I must work out what age the chicks are because that gives me an idea of when they will fledge. Normally, this is at 35 days.' From below us, apparently blind to our intrusion, a fulmar rises up and hangs in the breeze, its wings set and its body motionless as though it's a puppet on strings and there is marionettist high above us among the clouds. Ilka steps back. 'You know, for them this is hardly even windy,' she says, watching it slice away. 'They have incredible energy efficiency, designed perfectly for this weather.'

Ilka gestures to the scope and I fold my glasses away in my pocket before peering down into the lens. She has focused in on a flat ledge, some 10 feet above the sea, where an adult kittiwake is standing at the back of its nest with a bundle of pale downy grey tucked up beneath it. As I look at the chick, as it looks out over the sea with its little black eyes, Ilka tells me it will probably be 'ten or eleven days old, just with tail feathers starting to develop and primaries on its wings.' It's possible for kittiwakes to have up to three chicks and at one time this was relatively common, but Ilka thinks three-chick nests are generally now a thing of the past and sometimes only one will survive. 'If you look out there,' she says, pointing to the horizon, 'that at the moment would be okay, but if you have really stormy seas it completely affects breeding output.' Unlike guillemots or the puffins, kittiwakes are surface feeders, and if the water is too rough they can't sit out in the ocean bobbing for sand eels and their chicks will eventually starve.

As well as sand-eel stocks being plundered over the years by fishing boats, Ilka tells me that everything is moving north. 'It's current related,' she says, perching on a rock 5 feet away from the edge. 'Rising temperatures affect currents and copepods are passive.' She pauses and looks at me, then shrugs apologetically and says she guesses it's kind of complicated. 'Copepods, which is plankton. They drift, you know.' She raises her arms like little wings and says some of them do have tiny tentacles but it doesn't make much difference. 'What happens is that the current floats them in a certain direction and the sand eels must follow because they eat the copepods and the kittiwakes must follow also. It all goes north.'

Ilka smiles when I tell her about Billy Jolly's cursed scaly fish. 'You know, he's right,' she nods, sitting down and pushing her hands

deep into her jacket pockets. 'They're called pipefish. They are a
relative of the seahorse. They're not anymore available or have
somewhere disappeared, but it's awful because it's very spiny and
there is no nutritional value.' Ilka remembers being on St Kilda in
2010, the abandoned island 40 miles northwest of Uist, when kittiwakes spent the summer trying to feed pipefish to their chicks. 'Oh
gosh, it was just awful. Many many chicks did not fledge at all.'

I get back up to have another look at the nest. It is exactly as it
was, but just along the ledge, a couple of yards to the right, two adult
birds are changing places. 'They do that,' Ilka replies, when I tell
her what I can see. 'Kittiwakes are pretty equal. It's hard really to
distinguish between the male and female.' As I watch, hoping the
returning bird is about to feed its young, Ilka tells me that when
she's monitoring the database, people from right across the country
log on and upload all sorts of observations. 'Not numbers, just stories.
They want really to tell what they've seen and what they've experi-
enced. They just want to add their notes, but it never gets looked at.
I guess numbers, hard evidence, that unfortunately is science.'

Ilka takes her hat off and puts it on her lap. To our left, out over
the deep curling swell, the last of the fog has started to thin and
the sun is burning through. 'It is unfortunate that these stories don't
get looked at,' she says, running her fingers through her hair, 'but
we need to produce evidence for what is happening in order for
conservation bodies to act.' Beneath us, adult kittiwakes are setting
out in search of food and for a while we sit and watch in silence.
Above them, apparently unseen by the colony, a raven cruises the
cliffs in the hope of finding a chick that's strayed too far from the
nest. 'I don't know if we can turn this vast system of the ocean to
the kittiwake's favour,' Ilka says, turning from the raven to look
towards the horizon. 'This regime shift, as we call it, has been
happening now, in the North Sea, since the late 1980s and you

know, I think the time for a little bit of recycling and the pieces we try to change in our own lives has passed. It's time really to bring in laws for everyone.'

Ilka tells me she actually can't really imagine what it would mean if the kittiwakes that have forever returned to Marwick Head in spring no longer came. 'I don't see them as representing much,' she says, with her face resting in her hands. 'I think maybe the difference is that I wasn't raised here. I studied in Germany.' After graduating, Ilka went to New Zealand to carry out research on the sustainability of the Māori's traditional sooty shearwater hunt. 'It is important,' she tells me, 'to have an open mind. It is what they do and they really wanted us to come. This kind of practice will give you a completely different connection to the birds obviously, a spiritual one as well.' After being in New Zealand, Ilka had the opportunity to go on to study penguins or to work with seabirds in Britain. 'People think penguins are funny. They will put their own emotions onto them.' She stops for a moment, and glances up at me, seemingly trying to work out if I understand what she's saying, and then she turns back to look across at the cliffs. 'These birds, though, like the kittiwake, we see just the tiniest proportion of their lives. The rest of the year they're out there and God knows.' She shrugs and smiles, as though she's embarrassed for a moment and then goes on. 'There is a vast environment. We never see anything of them, you know. This is my fascination.' I stand up and wait for Ilka to pack her kit away but she tells me to go ahead. She has some more to do. Before I cut down through the fields, I turn and have a last look along the cliffs. Ilka has pulled her hat back on and is crouched down next to her scope, holding the photograph of the kittiwakes up to the light.

When John Cumming leans back, clasps his hands and raises them to his dry lips in thought, the little white dog he's been stroking on his lap lifts its head, opens one eye, looks at me coldly for a moment, as though to say fuck you, then goes back to sleep.

It wasn't until long after John's uncle died that he finally sat down and typed up his fishing diaries. 'That knowledge was intimate,' he tells me, running the back of his hand over the dog's soft belly as it rises and falls. 'Every day he was noting down vivid descriptions of the sea's colour. He wouldn't have known it but he would have been observing the plankton, which colours and thickens the water, and that will affect where the fish are, but he was also noting the gannets and the fulmars and the kittiwakes.' John suspects his uncle never knew the link between the sand eels and the plankton, but he remembers him talking about the connection between sand eels and fish. Lifting the coffee pot with unsteady hands, he pauses for a moment to pour himself another cup, then sits back in his chair and says that's how it was for men of his uncle's generation. 'They would only find fish by observation of nature and the sea.'

In the 1950s, when John was a boy on Burra, one of the smaller Shetland Islands – 'never more than half a mile across, sheltered on one side and with the Atlantic on the other' – he often went fishing in his uncle's boat, but when he left school he decided it wasn't for him and nor did he want to take on the family croft. Instead, he did what felt right and went to Aberdeen to study ceramics. Ever since, he's been drifting back and forth, between Shetland and Orkney, drawing, and sculpting, and writing in response to all the things that have inspired him and the traces they're leaving behind.

On John's left, five scorched seabirds, blackened in his kiln, as though they're the only survivors of a flock lost in an ocean fire, are spread out across a table. 'I made a raft of these as an installation in a show,' he says, passing me one. 'I wanted to capture the fragility and that seabird presence.' John's wife, Fiona, who has been sitting by the large studio window, picks another off the table. 'A disappearing race,' she says to me, with a grim smile as she runs her fingers over the rough clay. The bird in my hand is twisted and brittle, with a large hole in its breast and its back covered in charred fragments. 'I wanted to create something catastrophic,' John tells me, looking down at the rest of the raft on the table. 'There were 15 or 20 and I placed them all around the room.' When John leans over to take the bird back, the small dog jumps off his lap and heads for the stairs.

In some ways, as a crofter who worked the land, John feels that his father's relationship with birds was different to his uncle's. He didn't depend on them in quite the same way, but he'd grown up beneath a sky full of birdsong and he'd seen changes that saddened him. 'On the small island where we lived,' John tells me, 'there were geos where the kittiwakes would settle and the rocks echoed with their voices.' Outside, a pale wisp of cloud lifts and a beam of sunlight shines into the room, falling on John's desk. Behind it, sixteen skulls, spread out along a shelf, are lit up white: a curlew, oystercatchers and kittiwakes. Following my gaze, John looks across at them for a moment, then turns back and continues, 'When those kittiwakes returned, and their music began, that marked a time of year.'

On a chest in front of me, next to a small ditty box, five black books with an old map of Orkney on the front are stacked up on top of each other. A girl in the Stromness bookshop sold me the last copy in stock. 'I know the guy who made that book,' she told me. 'He was the best art teacher I ever had, but there was some sort of disagreement and he left.' She wasn't clear, but it was something about wanting the students to actually make art rather than learning about the theory. As I lingered in the shop doorway, waiting for the rain to pass, she told me – leaning across the counter, with her cheeks in her hands – that she's working on a novel of her own. There's enough people she reckons who come to Orkney, stay for a while and write about it, but what about those who are actually from the island, what about their voices?

'Was she a blonde girl?' Fiona asks. 'No,' John replies, shaking his head, 'that's a lovely thing to hear. That would have been Freja.' He leans forward to take the top book from the pile. 'The whole idea of *Working the Map*,' he says, passing it to me, was really about getting recognition for people who were aware years ago of changes. Before science was giving facts, your crofters and fishermen knew these things, but it didn't really count because it wasn't statistics.' John looks over as I leaf through, returning to a page I'd read the previous evening, where a local accordion player talks about being 'browt up ta believe that if you looked after the land, the land would look after you . . . you didna abuse the land, you cared for it.' I ask John if he thinks that for all the land needed love, there was a feeling that the sea could look after itself. He leans back in his chair and looks around the room. Outside, over the meadow that runs down to the beach, six or seven swallows wheel across the marigolds then turn and fly towards the window before casting up over the house.

'The sea when I was young,' John tells me, lifting his eyes slowly, 'was this vast thing. You know, it was just endless and cleansing. We thought it could cope with everything. We thought it was so big, and now in late life I'm thinking, my God, what have we done and will we get it back?' There is a moment of silence and I'm not sure what to say but he answers before I do. 'We can't. We can't get it back.' As well as remembering the ocean as ever-forgiving, John also feared it in a way that he thinks people no longer do. 'When

I went to sea with my uncle there was one cabin. You lay there together and you could hear the water go past.' As he speaks, he lifts his hand, and brings it slowly over the top of his head with his fingers outstretched. 'There was a closeness then and a real vulnerability. With today's super-trawlers and with all the technology, that's been lost.'

Under the telephone wires, down in the field, a rabbit hops beneath a fence and disappears into the long grass. After John's son finished school, he spent a while off the west coast in a Shetland trawler. 'He'd be fishing round Rockall,' John tells me, 'and he'd say to the crewmen, what's that bird there, and they would say, oh that's a maa.' Fiona stops him a second and tells me a maa is a gull. 'Aye, a gull,' he nods. 'So then he'd ask, well, what's that bird there and what's that one there, and they'd say, oh that's a maa.' He clears his throat and shakes his head. 'You see, I think, there's a huge cultural shift between the fishermen of my uncle's generation and the fishermen forever before that and the boys fishing today.'

In the 1970s and '80s, John remembers boats fishing for sand eel out of Shetland but he believes that's largely stopped now. 'They wanted them then for animal feed,' he tells me, 'but apparently now they're fishing them to run power stations.' Fiona laughs and says, 'It's the Danes apparently, but it's quite unreal, isn't it?' John tells me somebody was recently talking to him about it and he found it hard to believe so he rang up his son, who eventually became a marine biologist. 'The boy had a look and he said that's absolutely right, Dad, and at that time there seemed to be 60 ships in UK waters fishing sand eel.' Fiona stands up, puts the coffee cups on the tray and heads down out of the studio to the kitchen.

John thinks that when it all started to make sense to him was when he was on a boat, anchored overnight in the Minch with a number of other artists. 'I was looking at the kittiwakes on the

Shiants, there on the cliffs, and they were faring much better than the ones in the North Isles. I discovered later that there's a big breeding area for sand eels just beyond there and I started to realise then what was really happening to these surface feeders.' John remembers spending most of that trip sitting up on deck with a sketchbook, trying to capture the birds in flight. 'I'm not there on deck doing those pastels,' he says, pointing to two large drawings of kittiwakes hanging in the wind beneath an orange sky, 'but I'm studying movement.' He tells me he'll show me and then stands and lumbers stiffly across the room to the bookcase beyond his desk. 'Most of my work,' he says as he looks along the shelves, 'is couched in observation. It gives me a sense of when something's wrong. You'll be making a line and you get a real feeling that it's not right.' From among a row of hardbacks he pulls out two small pads and treads heavily, in his trainers, back across the carpet. 'These wee kittiwake books,' he says, passing them to me. 'I did some sketches when I was on my last trip on an old herring boat sailing from Orkney to Shetland and then made these using silverpoint when I got back.' John sits heavily. On every page he has sketched kittiwakes against the wind. Sometimes they are shadowy outlines, threes or fours, and sometimes there is a bird alone in great detail, dark wingtips and a darker eye.

The sun outside over the meadow has moved round and is casting John in a puddle of light. His hair is white and wispy and he has a shaving cut on his top lip, but his bright blue eyes aren't those of an old man at all. 'Lots of the people I recorded for *Working the Map* sat where you're sitting,' he tells me, smiling. 'I think in the past, science and the observations of the islanders just didn't meet. They haven't respected each other. I wanted that book to do that.' John remembers that when his father was alive he would often go and tell the scientists who would come to visit the island about the things he'd seen. 'They would patronise him. He was treated as well-meaning but not informed. Their attitude really towards the crofters was that they knew about these things and they were going to tell us.'

Fiona appears back at the top of the stairs and perches on a wooden chair by the desk. 'Would you mind just getting that urn?' John says, pointing past her. She stands and lifts a dark clay pot from a shelf behind the skulls. 'I spent a month walking Scapa beach to collect bird wrecks,' he tells me as she passes it to him. With shaking hands John lifts it up and peers inside. 'Have a look,' he says, holding it out for me. 'That's a kittiwake in there.' I tilt it into the sun and the light shines on the cremated remains of a seabird. Pushing my hand into the urn, I pick out a small piece of bone and rub it back and forth. In my sooty fingers, it crumbles into fragments and I let them fall among the ashes.

Eighty-five feet beneath, the Tyne flows fast and grey. It isn't raining but the city is swept beneath a rag of fog. In front of us, on the walkway, a dirty white sign is fixed to green steel, black letters asking whether we're 'in despair'. Below the question, there's a phone number. I look along the digits, wondering if anyone ever gets an answer machine or whether there's always a kind voice ready to try and convince you there's something left to live for.

Camera hanging round his neck, filmmaker Cain Scrimgeour leans over the rail, looking out towards Gateshead on the southern bank. His eyes are sunk deep into his skull and his beard is thick and red. 'That's where you can experience it just like the Farne Islands. You smell it before you see it.' As he speaks, Cain points in the direction of a large 1930s brick building, once a flour mill, now an art gallery, and one of the busiest nest sites at the world's most inland kittiwake colony. 'The birds are still on the roof. In the spring you'll get the brown speckled egg shells falling down and occasionally we get a mortality.' I peer through my binoculars, trying to find the right ledge, but the river curves round and I can't see the little white birds or anything dead on the walkway below.

Bumper to bumper, empty-eyed commuters are crawling over the bridge and Cain has to raise his voice over the rumble of impatient feet scratching at accelerator pedals. 'There's quite a lot that have moved off by now. It'll be mainly young that are still hanging round the nests. Probably just lazy greedy ones, like. They'll be waiting for the adult to come back and feed them.' Fifteen yards on, Cain stoops to pick up a feather and in doing so notices a bird perched on its nest just beyond the railings. Its head gyrates, back and forth, a kittiwake cartographer plotting the sky above the city. Cain crouches and I crouch next to him. I want to say the right thing but it's impossible to know what he's thinking. 'It's so clean. Look how clean it is.' It doesn't sound how I meant it to at all. As it gazes out across the water, it gives no indication it is aware of our presence. 'They aren't bothered,' Cain says cheerfully, 'because they've got this all day.' He gestures to the traffic and I look round. It has ground to a halt and the drivers in the cars behind us are peering down at the two boys squatting on the bridge. Cain starts laughing. 'I came along once and there was a very old guy, like a pensioner, trying to feed them bread. They don't want it. It was piled up round him on the ledge.' He passes me the china-white feather and I slip it into my pocket before following on across the bridge.

On both sides of the river, cyclists are pedalling to work. 'This would have been all industry and smoke once,' Cain tells me, looking to Newcastle on the northern bank, 'and when that ended is when the kittiwakes started. It was Dr John Coulson who spotted the first ones in the 1950s down at North Shields when he was crossing the

Tyne to go to a cricket game.' He stops and puts his hands on the rail. 'It was cricket or something. North Shields is quite far down, and they reckon they came first from Marsden Bay in South Shields, but they gradually bred in empty factories. Once they got knocked down it was no bother, they'd move into another one further upriver.'

Flying out from under the bridge, black wingtips cutting circles in the air, a kittiwake appears. Cain watches for a moment then says it's a juvenile, 'not massively confident, like'. I ask how long they live for and he tells me they've been known to make it to 15. Wings flapping, gracelessly, I half expect to see it fall at any moment into the cold morning, but it wheels round and flies back towards us. Cain stands ahead of me, looking down, seemingly watching for the point at which it disappears beneath our feet.

As we walk towards the end of the bridge, Cain is explaining that the kittiwakes at Tynemouth, where the river runs out into the North Sea, are usually two weeks behind the birds in the city due to the urban microclimate, but I'm distracted. A man in a green Peugot is rolling along next to us. He has one hand on the wheel and one hand down between his thighs, face contorted with a look of strained concentration. The gap to the car in front grows and the driver behind him bangs her horn. He throws a phone up onto the dashboard and his car jerks forwards.

'You see that?' Cain points up to one of the bridge's granite towers. 'It's a stack nest. They don't do them anywhere else in the world. It's one of the things that makes the Tyne colony unique. At the sea cliffs, the winter weather would just knock them down. They have to make a new nest there each year.' On a narrow window ledge, high above us, a muddy mix of grass and weed is piled up, one layer on top of another, baked hard by the Tyneside sun. I begin counting out loud but get to a different number twice. 'I make it ten year,' Cain cuts in, 'at least ten year and I can't remember which factory it was but one of them had these massive stack nests.' He starts telling me about photographs he took just before the factory was pulled down and how the dominant kittiwakes always get the pick of the best nests, but he's drowned out by an ambulance wailing across the bridge.

We cut right, across Pilgrim Street, winding our way down into the sticky bowels of the city and Cain tells me about discovering birds. 'I think it came from my dad and grandad. They weren't wildlife folk but they were more like, you know, countryside folk. They grew up from mining villages, my dad's family, so they were into birdnesting and they had hawks.' When Cain was a child he was surrounded by men who spent hours fishing but he has been wondering, lately, whether any of them were interested in catching fish at all or if they just wanted to be able to spend time standing in the river without looking strange. On our left, All Saints Church, Doric columns and a steeple atop a square bell tower, sits on a strip of brown grass. At the bottom of the grand stone steps, a pigeon is stooped over, eyes fluttering to a close, head lolling and a balding wing trailing on the concrete. 'It was seeing a crested grebe that really sparked us. It was right by where I used to live, in Whitley Bay, and I just thought how could something so different and stunning be so close to my home and I didn't even know it was there and that was it, like.' But even though it was only 15 years ago things were very different then. Cain thinks there were probably other people at school just like him, but none of them would tell anyone because 'we'd get the shit kicked out of us'. It was only when he went to university that he remembers taking his binoculars out of his bag and finally daring to wear them round his neck.

When he's not working with a camera, Cain leads walks around the city showing people the life they share the streets with. 'Kittiwakes is a perfect one,' he reckons, 'because people just go straight past, but once they've seen them and they learn that other than being here they spend their whole time at sea, I don't think they ever really look at them again in the same way.'

The steep pavement, down Akenside Hill, where a row of butchers once stood and a parking meter stands now, levels out into Newcastle Quayside. At night it is a place for dancing and sitting on kerbs eating kebabs, but all summer long the air peals with the kittiwakes' cries. Cain holds his hand up and draws his finger in a wide circle above his head. 'It's deafening in spring. With the architecture it just echoes.' There was a time when every grey stone ledge, running across the towering Victorian buildings, was covered in kittiwakes, but in the last decade as the colony grew, businesses started putting up netting in an effort to stop the birds nesting. 'Can you see that one tangled up there?' I follow the direction he's pointing in but I can't find it. 'This was one of the problem buildings. See, in between the two turrets there's a bit of white.' Looking through my binoculars, I run my eyes up the stone. I ask Cain if it's alive but he shakes his head. 'When they're alive they flap around trying to get free.' Beyond the highest window, where the dark slate roof begins, I see the white weight hanging, head tucked up beneath its body.

On our left under a neon sign, a narrow door between boarded-up windows bursts open. 'Watch your feet!' We move aside to let a lady, yellow-toothed and smiling, throw a bucket of soapy water into the gutter. I tell Cain, as we wander between the buildings back towards the river, that I read something about kittiwakes being the souls of dead children. He nods. 'That's the one. I think it's the sound, and when you look into their eyes it's like staring into the ocean.' A streetlamp is guttering and Cain stops in front of it. 'See, they've even taken to nesting there. The light always stays on because they're sat on top of the sensor on a pile of mud and seaweed.' We stand, watching people come and go, heads down, ignoring the bird above, while it looks out across the Tyne, ignoring the people beneath.

It is mid-morning and the sun is casting swirls of old gold through the water. As we walk, Cain tells me the Tyne has changed. When he was a child, it ran dark and dead, but in recent years otters have been spotted where the shadows of hollowed-out factories once loomed, and in autumn, if you wait long enough, you sometimes see salmon running through the city to spawn. Where the path along the river cuts beneath the bridge, he stops and points to one of the great stone supports. 'You can't tell anyone yet, but we're going to put a kittiwake on that. If you think about it, if you want a new audience you've got to tailor what you're doing.' I nod and his voice rises with excitement. 'I want people to smile at the kitti-wakes, not just think of them as those birds that shit on us when we go out. We've got an artist lined up, he does murals of endangered British species and he's going to do a massive kittiwake on that.' Beneath the green steel struts, torn netting still hangs down across the water. It is left over from when the entire underside of the bridge was covered up to keep the birds away. Twenty yards across, two kittiwakes bob, black eyes turned skywards as though wondering if it is time for them to fly back to sea for the winter.

Putting down roots

No one invents an absence:
cadmium yellow, duckweed, the capercaillie
– see how the hand we would name restrains itself
till all our stories end in monochrome.

The path through the meadow
reaching no logical end;
nothing but colour: bedstraw and ladies' mantle;
nothing sequential; nothing as chapter
and verse.

John Burnside, 'Si Dieu n'existait pas', *Gift Songs*, 2007

In the 1830s, at the age of 29, James Giles, the Scottish landscape painter, travelled north to Aberdeenshire. Ever since his father died, when he was just 13, he had been supporting his family through his art, but Giles was on different business. His friend, William MacGillivray, the first naturalist to distinguish between the hooded and carrion crow, had asked him to make notes on the capercaillie for his forthcoming study, *The Natural History of Deeside*.

At Old Balmoral, Giles called in to see Sir Robert Gordon, a keen sportsman who eventually choked to death on a fish bone. Gordon was fascinated by local birds and he wanted to show Giles a painting. The watercolour, hanging on a castle wall, depicted the shooting of two capercaillie in Ballochbuie Forest, beneath the foothills of Lochnagar. On the back of the frame, Giles found some words in the handwriting of the unknown artist, which noted that the birds were encountered during a successful hunting trip, the morning after a Highland wedding in 1785.

Giles wrote that Sir Robert was confident these were the last native capercaillie ever seen in Scotland. The notes didn't make it into MacGillivray's book, the castle was torn down just 20 years later by a new German owner, and the painting was lost. It might turn up yet, in with the china dogs and the hi-fis at the local car boot, yours for forty quid some sunny Sunday.

Through the blaeberry, in the dying light, they come towards me, their guttering growing louder and louder until they are gathered all around. At half eight, the first one rises to roost, a rush of wings through the chill April air. Five minutes later another follows and then another, until there are six cocks calling in the tops of the pines. At eleven, when I wake, the forest is black and my bladder is full. Like Ewan told me, I scrape away the moss in the corner of the hide and piss into the little well. 'It's that or a bottle,' he'd said five hours earlier before he walked back to his truck. 'If you go out when it's dark, that'll be them gone.' In my sleeping bag, I lie quiet, eyes closed, listening to the midnight groans of lorries on the A9.

In 1836, Larry Banville, a Wexford-born gamekeeper, boarded a boat to Sweden. His master, Thomas Buxton, who was back in Norfolk, was a radical man. When Buxton died, at the age of 58, he had made a great fortune, led the battle for slavery to be abolished throughout the British Empire, served as the founding chairman of the RSPCA, and reintroduced capercaillie. Banville's journey was an elaborate thank you. The summer before Banville set out, Buxton had been to stay at Taymouth Castle, Lord Breadalbane's Perthshire estate. Buxton's distant cousin, Llewelyn Lloyd, a man who cared very little for work and very much for Scandinavian salmon rivers, had previously gifted Buxton a pair of capercaillie that had been caught among Sweden's great pines. The gift wasn't a successful one. The hen and her chicks died in the sun and the cock, which was released into the woods at Cromer on the North Norfolk coast, was shot by an over-zealous sportsman. In truth, the habitat around Buxton's estate was relatively unsuitable for capercaillie and reintroducing them in England, where they'd been extinct since the 1660s, was probably an impossibility, but Breadalbane's land had greater potential.

In 1837, the *Dundee Courier* reported that Banville had passed through town, on his way to Taymouth Castle, with 'twenty-eight birds of the cock of the wood species collected throughout Sweden'. For metropolitan Dundonians who knew little of the natural world, the report notes that while capercaillie have long since disappeared from Scotland, 'in some countries on the continent, especially Sweden, it continues to live free from molestation'. The release was a success and Larry returned south with 5 pounds from Breadalbane in his pocket and a message for Buxton that the Scottish aristocrat had no idea how he could ever repay his kindness.

When larks are still lying low, during the coldest part of the night, the capper start singing in the trees above me, popping and gurgling before the sun comes up. The previous morning, 3 miles down the

Spey, the hens visited the south lek and there's every chance, Ewan told me, when we were putting up the hide, that copulation might be today. The zip rips through the stillness – everything is loud when you're trying to do it quietly. The mesh panels have one-way vision, camouflaged to a capper looking in from the outside and supposedly perfectly clear when looking out from within. I sit shivering, looking out over the lek, but it's still too dark.

Somewhere behind me, just after the blackbirds have begun, when dull light is seeping slowly into the sky, the first capper drops from its roost, landing heavily on the forest floor. With a pen, I make a mark on a sheet of paper to try and work out roughly where it is and another when the next one drops and then another, until all six are down among the trees. It's an hour before I can tell where darkness ends and the forest floor begins, and half an hour after that until I see a flash of rich blue. Twice I think it's not a capercaillie at all before twice deciding it must be and then it turns its head, a large powdery pale beak standing out against the heather. According to the latest estimate, the bird perched 40 yards away from me is one of just 1,200 that remain in the remnants of Scotland's pine forests, and there are very few people who don't believe that their second extinction has almost come.

Out in the distance, on a hummock, where light is shining down through the canopy, the dark, white-flecked tail feathers of a displaying capper are fanned out behind a tree. While I eat my gingerbread loaf, the bird stands poised and then canters forwards, head held high and voice rising. Until eight, the dominant cock clatters through the understorey, rasping and clicking and stopping only briefly to leap into the air, wings beating, and puffed-up neck shimmering in the sun.

By the time Ewan appears through the trees, long strides with his hands in his pockets, I've already packed up most of the hide. As he crouches to pull up the pegs for the guy lines, counting them under his breath as he drops them into the bag, he tells me that 'the hens will have been roundabout somewhere, hearing it all, but you just cannae predict when they'll come'.

Ewan Archer was 16 years old when he started as a trainee keeper at Seafield: an Aviemore boy working on the 90,000-acre estate that his home town was built on. Thirty years later, he has three men under him and his ginger beard is white at the edges. Back when the estate factor sent Ewan down to Beauly to get his first set of tweed breeks made, it had already been a decade since shooting capercaillie had been banned due to dwindling numbers, but at that point he still remembers seeing them so frequently they were hardly comment-worthy. Over the past few years though, even on Seafield, where there is a bigger capercaillie population than on any other estate in Scotland, Ewan has started to think that they don't have long left.

As we walk by the stance where the forlorn bird was perching, Ewan kicks at a flattened hollow. 'This is where I was the other night. What I do is I pull up the moss to insulate the gaps in the tent.' Ten yards further on, he turns, laughing, and says he maybe should have mentioned that one when we were building the hide. Yesterday was the first time Ewan had slept in his own bed after being out for almost a week in order to monitor every lek across the estate. He tells me that the old boys thought nothing of sleeping out among the trees, but although he loves it, and 'it's all part of it', he's so tired by the end of a six-night stint that he can hardly speak.

Near the patch where the dominant capper was displaying, Ewan reaches down, picks a tightly formed pellet out of the dust and passes it up to me. He runs his fingers through the red earth. 'See the way that's broken there. I would say that would be a capper. Very often in here you get a dustbowl where they're preening, you can see they'll have been on these sticks and there's tonnes of capper shit.' He rolls another pellet between his fingers, then pulls it apart. 'Because birds don't pass urine,' he continues, 'the white is urea and the vegetation is Scots pine. Ninety per cent of the capper's diet is pine needles but the nutritional value is terrible, so this time of year hens will be eating a lot of moss crop because it's so full of protein. That's her getting herself into condition to start producing eggs.'

While all sorts of people will tell you that Scotland's capper population sits at just over 1,000, Ewan thinks that due to productivity being so poor over the past few years, things are actually far worse than that. 'It's been like 0.2 chicks per hen we're seeing and personally I'd say the numbers now are more like low hundreds.' Where the ground rises up to the track that Ewan came in on, he turns to look back at the lek site. In the past few months, he's commissioned a study to come up with a revised population estimate in the hope that it might make people realise the truth of the situation. 'Believe me,' he says, wringing his hands, 'we need the shit to hit the fan when those numbers come out.'

At Seafield, Ewan thinks they are still doing almost everything they can for capercaillie. Over 30,000 acres of the estate is pine forest, making up 9 per cent of the country's total, and they've just taken on another gamekeeper to focus solely on capper, but they can't exist as an island. 'With our neighbours,' he tells me, when we get back to the truck, 'we need to be feeding them capercaillie and they should be feeding us capercaillie. Young cocks tend to spend their lives in the same area but there's a great deal of dispersal with females. As they become up to a year old, they can go twenty kilometres.' In some cases there's no pine woodland for the birds to go to, but Ewan is just as concerned about the lack of predator control happening beyond the estate boundary. He takes my bag, throws it into the back, then gets into the driver's seat. 'We've got RSPB Abernethy next door, which they bought from Seafield as a flagship capercaillie reserve, but their capper are on their arse.' He speaks slowly, picking over every word. 'Not a crow or a fox will they kill. For us, as neighbours, not only do we get steady foxes and crows coming in but we don't have anywhere for our capercaillie to mix. As a neighbour, it's a shocker.' Oily dark hair and small black eyes, Ewan's terrier, Titch, nine years old now and no longer interested in working, jumps up and sits on my rucksack, staring out through the windscreen into the trees as we wind our way down the track.

Just over two years ago, the BBC got in touch to ask whether they could film *Springwatch* at Seafield. 'They knew we had the biggest lek in the country,' Ewan tells me, 'and lots to show them: white-tailed eagles, golden eagles, two nests of ospreys, but we refused them. The producer phoned, wanting to know why and I told him. I said your presenters are doing their best to shut us down. I said we'd discussed it as an estate and they weren't coming.' Overhead, against the sun, a growking raven skims the wind then drifts away above the pines. A couple of weeks after the producer called, one of the cameramen got in touch and Ewan felt he had to apologise. 'I had to say to him, look we've talked about this as an estate and they aren't coming. That man, he'd have me sacked.' Ewan looks at me and shrugs. 'The cameraman, he said to me I'm fucking sick of this. He said I'm fucking sick of it.'

An old man with a shaking dog, red checked shirt and grey face beneath a cowboy hat, stands on the side of the track, no teeth when he talks. 'Cracking Morning isn't it, what are you up to today?' Ewan tells him we were monitoring a capper lek last night and it's good news. 'Nine cocks all together.' The man grunts in a mid-pitched, impressed sort of a way. 'I sometimes see them in the morning right enough, but now tell me, you don't often see a hen capper, do you?' Ewan nods. 'It's a big problem this year. There are cocks at the leks but very few hens.' The man shakes his head and stands for a moment. 'I'll tell you, it's a very sensitive issue but what about these badgers? Jesus, I never seen so many badgers. You cannae comment on that, I suppose.' Ewan laughs and tells him he absolutely can and it's one of the things that's driving the big decline in capper. The dog shivers on the ground and the man is wide-eyed with the nonsense of it all. 'You'd think after it was on yon, was it *Countryfile*, it was a badger nipping, was it the avocet's eggs? You'd think folk wid taken a changed tune. You see a lot of foxes about too. Then again, they all have to live.' Ewan says he thinks it's a political thing really. 'No politician will tackle it. No politician wants to talk about badgers because it's not a vote winner. It's a vote loser.' The man shudders. 'I can't see the appeal myself. They're vicious buggers. If you get one in with the hens or something like that . . .' He breaks off. 'Oh Jesus, but I don't suppose you can do anything?' Ewan shakes his head and says he's got to run within the law and just hopes that at some point someone comes along with the balls to change things. 'It's embarrassing right enough,' he admits, when someone calls up to tell him something's been at their hens and there's nothing he can do about it. 'There's only one vote winner nowadays,' the man replies, 'and that's to bloody bribe the feckless and the work-shy.' Ewan laughs, and the old boy continues, 'That's the way it's working, like. Build up a voter base with them that's getting handouts.' Titch has clambered down onto my lap and is straining her little neck to see the white dog. 'A Jack Russell?' Ewan asks. 'That's what they told me anyway,' the old boy replies. The sad-eyed animal looks up at his master. 'Right, wee man,' he says to the dog, 'time to get you home and get the coffee on,' then asks, just before we drive away, what we think should be done about China. 'It's time somebody did something anyway.' As we turn out onto the road at the bottom of the track, I ask Ewan who the man was. 'No idea, never met the man in my life,' he replies, 'but I'd say he was from Aberdeen.'

Where the bog cotton hangs heavy with dew and greylag geese are tucked up along the burn, we wind our way through the fields. Ahead of us, thousands of acres of woodland run out onto heather and further still the ground rises up to become the grey, treeless face of the Monadhliath mountains where the tops are swept in the last of the winter snow. 'Would you mind getting this gate?' Ewan brings the truck to a stop and passes me the keys. 'Just leave it open if you can't see any cattle.' Around the fringes of the meadow, rabbits are grazing in the warmth of the morning sun. When I get back in, Ewan tells me that there used to be hundreds of them in every field and they kept the wild cats fed but 'there's no such thing now, they've all interbred with domestic tabbies that come up from the villages'.

Just a couple of decades ago, the red deer on some parts of the Seafield Estate numbered 20 beasts per square kilometre and every year paying guests would come to stalk stags. Ewan remembers it being a fantastic business that kept the lodge going, but it became clear that the numbers were having a serious impact on woodland and the estate was hit with a Section 7. 'What they said is you need to take your deer down to five per square kilometre and if you don't, we'll do it for you and you'll pick up the bill.' Where the track climbs up into the trees, Ewan tells me it was something they realised that the best lawyers in Europe couldn't fight, and while they didn't go into it smiling, 20 years on nobody denies the benefit of woodland expansion. Even though they got the numbers right down, it's an ongoing job, and back in January they shot 230 hinds as part of the annual cull. Ewan admits that, in truth, the older he gets, the harder he finds it to shoot that many animals. He winds the window down and we drive in silence. 'It's a feeling more than an ethical thing,' he says after a while. He nods, 'Aye, I would struggle to go on with it if I didn't think I was doing it for capercaillie. I appreciate the woodland but I'm going to struggle in this job if we lose the capper. To me, it's part of our heritage and we're in great jeopardy of losing a lot of that.'

Beyond an old dyke, where young pines are coming up, Ewan turns the truck onto the grass and stalls in a rut while turning. He turns the key again, stalls again, then leaves it a moment. In front of us, running along the road, there's a fence with wooden slats fixed to it. 'That's part of the problem,' he says, pointing at it. 'There's 24,000 metres of deer fencing I've marked like that. When you mark it, the capper see it as a barrier and fly over it, and 30,000 metres we've removed. In the past, you'd just find dead capper at the bottom of the fences.' Ewan starts the truck again and we head back the way we came. There is a rattling noise coming from the back and he holds his hand up, telling me to be quiet for a moment so he can listen. When it fades he continues. 'The fences are a big part of it, but it's the whole picture. We've felled their habitat, they've been disturbed, we've planted the wrong trees that provide them with no food, and they've been predated. We've all had a hand in it and I feel hugely responsible for the capper that remain.'

When capercaillie were at their best on Seafield, pine martens had been entirely eradicated and Ewan knows old men in the area who remember the first of them returning in the early 1980s. 'People were seeing these things and they were saying what the hell is that and there's no doubt about it, there's a massive link between capper decline and pine martens.' Over the years, Ewan tells me that the Scottish government has been given plenty of pictures of pine martens coming out of capper nests with eggs in their mouths, but they still maintain that there isn't enough evidence to allow people to shoot them. 'We all know that the pine martens are expanding. It's not rare. It's not endangered and it's having a massive impact.' Ewan shakes his head when I suggest we can't really go back to the heavy-handedness that saw pine martens eradicated in the first place. 'That's not even entering the discussion. Pine martens and badgers, buzzards, goshawks, sparrowhawks, foxes, they're just as much a part of this woodland as that pine tree,' he gestures out the window as he speaks, 'or that blaeberry or that juniper bush.' He stops a moment as though deciding which direction to go in. 'All I'm asking for is an entire view of what we've got on this estate to allow us to manage predators within reason. So what I'd be talking about would be lethal control at certain times of the year to allow capper to flourish. I'm not talking eradication. I would never suggest that. Nobody is discussing eradication.'

Back down on the meadow, the geese have flown. 'It's like our badgers,' Ewan begins again, when I get in after shutting the gate. 'We've got a badger sett in most of the woodlands on the estate, but how many do we need? Our badgers now outnumber our foxes fifteen to one. Do we need ten badger setts in some of our woodlands, do we need to go out at night and see forty? We have to make a move soon or we're scuppered.' When Ewan is lying out at the capper leks he hardly sleeps. 'But when I'm back in my bed,' he tells me, 'I often lie awake too, thinking about the capper. In order to actually get any sleep, I need to know that I'll be able to say I tried as hard as I could and I fought our corner for some modernisation in the law, for a bit of radical thinking.' At Carrbridge, two old couples are standing in the road taking pictures of the humpbacked crossing over the River Dulnain. 'It was built in 1717,' Ewan says as we swerve round them, 'so people could get to church when the water was up. It's a lovely thing.'

Hanging on a line from the garage to the back door, Ewan's sleeping bag is drying in the breeze. He gestures to a bench in the sun and I sit as he walks round to the back of his truck and lifts the tailgate. 'Button! Button!' He calls the name twice and a little terrier, Titch's daughter, scrambles out. 'It can all just get a wee bit heavy duty if the dogs are stuck in there together,' he tells me as he walks across the gravel. Ewan sits at the other end of the bench and pulls the dog onto his lap. 'Bolted her first fox last Monday,' he says proudly, lifting the little animal into the air. 'Never been underground in her life. Nearly 400 earths we have to check now, every spring, for signs of foxes because of the excavations of badgers. There's no other way, can't snare because of capper and can't shoot because of the trees, so we have to use terriers. I let her off and she flew straight down.' Karen Archer, peroxide hair and 10 years younger than Ewan, opens the back door and comes out with our sausage rolls and a pot of tea. Button dashes over and walks beneath the tray on her hind legs with her nose in the air. 'I was just saying this morning, Karen, to Patrick, that if I didn't have capercaillie, how would I get out of bed every morning in the winter and cull deer the way I do?' Karen pours herself a cup of tea and sits down across from us. 'Well, I'd have to kick you out.' Ewan shakes his head. 'The reason I do all that is for the capercaillie, Karen.' She nods. 'I know you do.' He picks up Button and walks across the gravel to meet the mail van coming up the drive. 'First fox last Monday,' Ewan says to the postman as he passes him his letters. 'She's going to be a cracking dog.'

Perched on the weathervane above the clock tower on the large grey building, two rooks sit squawking. On a fine day the noise would fill Findhorn's narrow streets, but down on the pier, in the biting, growling gloom, the birds are barely audible. Across the bay, where every spring millions of salmon used to run, deflated buoys rise into view as the tide drops away around black sandbars. A rough mile over the estuary, a pine forest is fading as the light goes, the trees no longer individuals but a great dark green mass. Capercaillie stalked that darkness, just 30 years ago, but they exist only as memories now, lekking in the minds of the old folks.

Down on the rocks, where rotting seaweed meets the water's edge, a toddler and her father skim stones. He stoops, holding her little white hand, trying to show her how you make them bounce over the wake. Standing, he twists his back uncomfortably and then skims one of his own, watching as it cuts out into the grey before being swallowed by a breaking wave.

In his moment of absence, the toddler is gone, running away up the beach, froggy boots splashing through the mud. He shouts her name and there is only anger in his voice, and then he takes off after her and I see that he is limping, right leg trailing after the left. A hundred yards on, the girl stops and crouches, trying to pick up a buoy. By the time her father reaches her, she has dislodged it and lifts it as high as she can manage before letting it fall forwards onto the water. Scooping her up, the man throws her over his shoulder and walks towards the houses, him in silence and her with tears streaming down her face.

'You're finding your way all right.' Hood over her head and scarf up to her lips, I don't recognise her at first. When she says it again, smile fading and confusion running into her voice, I realise the wrapped-up girl with the pigeon-toed terrier, standing on the pier behind me, is the one from the Crown and Anchor who took my bag up to my room. The Staffordshire's eyes meet mine, and the girl tells me how gentle he can be, before turning and walking towards the harbour where the halyards are clanging on the masts. It was the way I was going to go, but I head right instead, up into the cold shelter of the streets, astragal windows with curtains drawn against the weather. At the end of the main road, beyond a crumbling boathouse, a grass bank slopes down to the sea. Out somewhere on the saltings, a wildfowler's shot rings out and the peace of the estuary is broken. All across the bay, curlew, lapwings and dunlin lift, crying shadows twisting in the last of the light.

My fork pierces the breadcrumbs and thin grey liquid runs across the plate. 'She's a trapeze artist, you know.' The mother of the girl on the pier who took my bag up to my room is standing over me, back to the fire, telling me everything I might want to know about Ellie. 'Not a trapeze artist but she dances, you know, on the ribbons. That was in France before she came back. She had a boyfriend. He was a trapeze artist, roofer now.' Ellie appears from the kitchen with a man I take to be her father and the lady hurries away. When I break through the shell I find that the scampi tail is shrunken and flaccid, too long entombed in the bottom of the freezer – I should have eaten it whole.

The big man is standing at the till, prodding the screen. 'Show me again, Ellie, I want to see if there's been any more slagging.' The girl wanders round patiently, while telling her dad it's only been an hour since they last looked. She clicks the computer into life and starts scrolling, 'No,' she replies, 'just the one review saying it was too expensive.' On his way to the door, he repeats the charge twice, and then stops with his hand on the handle. 'If it's only him by seven, pull the bolt across. I'm away for a Chinese.' When he's gone, she turns to look at me and then disappears into the kitchen, and I'm left alone with the sound of the fire hissing in the grate and a pack of whining winds begging at the window.

Twenty minutes later, as I'm drinking the last of my second beer, a soft knocking sounds and Ellie unbolts. She tells whoever it is that she 'won't be a moment', then disappears into the kitchen. 'It's that nice English couple,' I hear her say, before she re-emerges and heads to the door. The pair shuffle through and take off their gloves before holding their hands over the coals. When they sit they order 'the same as last week. It was lovely, with the same wine too, please.' Ellie fetches a bottle from behind the bar and places it down on the table before wandering to the kitchen. 'We should still do some planning,' the woman says to the man sitting across from her, 'catering at least'. The man lifts the bottle and lowers his voice. 'What about serving this?' The woman starts laughing. 'I don't know where the fuck you even buy wine like that.'

There's still been nothing from Alan, and if I have a third beer I won't sleep. On my way upstairs, I notice, hammered to the wall, photographs of village life from three or four generations ago. In one, men with woollen caps and chiselled faces haul a net of salmon up the beach, and in another, the carcass of a fishing boat, half sunk in the sand, rots away. At the end, in a chipped wooden frame, there is a photo of the Crown and Anchor; all the doors and windows have been thrown open and there are smiling faces in every one.

I knock three times and then shuffle back down the garden. Alan Watson Featherstone emailed in the night. I was to meet him, he wrote, at his house on the edge of the Field of Dreams. After a minute, I go to knock again, but as I reach up, a lady appears, unlocking the door and opening it just enough that she can poke her head round. Early fifties with silver-flecked chestnut hair, she eyes me with solemn indifference, giving no indication she was expecting anyone at all. 'I'm here for Alan, about the capercaillie.' She nods. 'He needs to finish his tea.'

The air is thick with fog, and in the sky above the Findhorn Foundation a seam of pinken gold shines through. Alan's house, timber clad, with stained-glass charms hanging in the windows, glows warm. The lady walks back in and then reappears a minute later and pushes the door wide open. Alan is doubled over, his hands tying his laces, and his face turned towards me but with eyes that look everywhere else. He is a little man with big white hair flowing from beneath a blue knitted hat. He stands, kisses the lady on the mouth and then holds his arms out for her to help him put his jacket on.

Alan scrapes down the path with little steps while doing up his zip and I walk slowly behind him. It is 47 years, he tells me, since he washed up at Findhorn. He was 24 years old then and was trying to find some sort of purpose after becoming aware, while doing yoga, that he was neither his body, his feelings, nor his emotions. 'I had this experience,' he continues, as we wander among the cabins, some large and some small but every one of them different, 'where I felt my consciousness withdrawing from parts of my body, and suddenly I realised I'm not this body but I live in this body.' I ask him what he discovered he was and his darting eyes catch me with a cold glance before he turns his head away. I have come to talk to him about Trees for Life, a charity he founded to restore ancient Caledonian pine forest after realising that without it there would be no future for creatures like the capercaillie. But it all started long before that with a caravan, a garden, and a Canadian mystic, dead last spring at 100 years old, who spoke to the nature gods.

'BE STILL AND KNOW.' A wooden gate hangs between two pine posts, with the blackened words carved along the top and over in the corner, cream roof and faded blue sides, an old caravan stands on crumbling bricks. 'This is where it all began,' Alan tells me, 'when Peter and Eileen Caddy and their friend Dorothy Maclean, as well as the three Caddy boys, pulled up here in 1962. It was just sand dune then and they lived in the caravan for several years.'

I push the gate open and step between the flowerbeds. When Peter and Eileen arrived they were scraping by on the dole, 8 pounds a week, after losing their jobs. A vegetable garden was a necessity but the conditions were impossible until Dorothy was contacted, while meditating, by angels who made it clear to her that to have any success, she would need to first attune to the inner essence of whatever it was she was trying to grow. She began, on the basis that it was one of her favourites, with the garden pea. To her astonishment, the pea told her that we are 'all great beings of light and we aren't using our capabilities'. Through being able to communicate with vegetables, Dorothy was able to understand what it was they needed and the garden went from being a failure to an extraordinary success, with the notable production of 40-pound cabbages.

In time, word of this esoteric spirituality spread and fledgling

New Agers from across the world began to drift to Findhorn in the hope of finding inner peace in living differently. Unintentionally, a conversation with peas was set to give rise to what has become one of the most successful communes the world has known. There are very few people left at Findhorn whose hands have worked the soil alongside the three founders, and Alan now provides a sort of spiritual link to where it all started. Down the years, the Foundation has become increasingly spiritually eclectic, but so much of what Dorothy and the Caddys drew on, and the way they believed nature needs conscious love, has informed Alan's own efforts to restore Scotland's lost capercaillie habitat.

The chrome handle on the caravan door is flecked with rust, and moss is growing on the hinges. When Alan is down at the other end by the raised beds, I press my face against one of the windows. A rip runs down the nylon curtain, but all I can see is a patch of brown floor with a dark stain spilling across it.

Beyond the rhododendrons, past the last of the solar panel-topped houses, we cut along the fringe of a Scots pine plantation planted in the 1940s, just after the war, and I ask Alan if any of it is true, if love really makes the trees grow. He is ahead of me on a narrow path and I can't see his face. 'All life responds to love,' he replies, before telling me, by way of evidence, that he has never had any success growing spring onions because he doesn't like the taste and has therefore never embraced them in his heart. From among the trunks, a smiling lady emerges, singing a hymn I don't recognise. 'Lovely to hear your voice,' Alan says as she passes. She turns her head to sing at him, smiling face unchanged, before carrying on ahead of us. Back in 1990, Alan continues telling me, some years before he announced, standing in front of hundreds of people at

Findhorn, that he was going to devote his life to restoring wilderness, he noticed a seedling growing on a hummock. The tree was at Glen Affric, which is one of the first places he started erecting fences to keep red deer out. Like a loving father, he returned every summer to take a photograph of himself next to the pine and often took others with him to admire its beauty. 'Then on the twentieth anniversary,' he tells me, his voice blowing up like a priest who is about to show a doubting boy some incontrovertible proof of God, 'I went with someone who had a laser hypsometer, which we used to take height readings of all the trees.' The question hangs in the air and it would be callous not to ask. Alan nods piously, 'Exactly. The one that got all my love was the biggest. You see, you and I, all of us, we are beings of love and light and one of the abilities we have as humans is to consciously focus that light and that energy.'

After we've left the plantation and have ducked down into a sheltered hollow, beneath the low-hanging branches of an open grown granny pine, Alan recalls that it wasn't long after the loved seedling sprouted that he saw a capercaillie for the last time. It was 1993 or maybe '94 and he was driving up a forest track in a clapped-out Leyland minibus when a lekking cock bird leapt onto the windscreen and tried to attack him. After winding the windows up, he reached for his Polaroid, and now often reaches for those old photographs. The following year some of his volunteers thought they saw the same bird again, but after that and ever since, despite his efforts, capercaillie have been gone from Glen Affric.

Like an elfin king on a Scots pine throne, Alan perches on a knot bulging out of the wizened ochre trunk and I stand before him, leaning against a branch that sweeps over the needle-strewn ground. Above us, limbs criss-cross, entangled at jaunty angles. In another age, it would have been a fine place for a capercaillie hen to spend a summer's night with her chicks, high above prowling predators.

'Look at this,' Alan says, head held skywards. 'It's beautiful, look at how it's exploring space three-dimensionally. Look at its freedom of expression.' He stands as he speaks, pushing his chest out and rotating his arms to mirror the movement of the tree. 'Branches tilted to the light,' he continues, before he slumps his shoulders forwards and points back the way we came, through a gap in the needles. 'You see what we've done?' – his voice is accusatory – 'we've forced the trees to stand straight. The pines in that plantation are regimental, unable to express themselves, and because of how closely they're planted there's no room for any limbs to grow outwards. That sort of forestry, which we see across Scotland, provides little benefit for birds like the capercaillie.' In a field to our right, a high-land pony whinnies, its thick winter coat heavy with mud. An old lady is coming down the track towards it with a wheelbarrow of hay. 'It's the arrogance of humanity,' Alan continues, as I watch her trip and stumble over the ground, 'that's why you've got those plantations. People think we can do better than nature so we're going to plant trees, but nature knows best. Nature has taken millions of years to evolve its interconnectedness, things we're only just beginning to grasp.' Someone used to ride that pony, Alan tells me, but then they went to California and he's getting fat now.

We clamber over the limbs and carry on towards the sea. The temperature is falling and greylag geese, seemingly disturbed by something or someone on inland barley fields, are flying across the sky, back to the mudflats where they'll roost for the night. Where the ground beneath our feet begins to shift from mud to sand, Alan tells me people laugh when they hear he has only planted two million trees, but they laugh because they don't know. His guiding ecological principle, he explains, has always been to 'start from areas of strength because where the ecosystem is closest to its original state, relationships that we'll probably never understand already exist.' There has to be a degree, he accepts, of planting trees where there are great gaps, but there must also, if there is to be any hope for the capercaillie, be a shift of focus towards protecting natural forest fringes with well-marked fencing. 'In the meantime,' he continues, 'we must get numbers of deer and sheep down, the two herbivores that have kept would-be capercaillie habitat looking like a contoured lawn for hundreds of years.' It is only when we pursue both of these aims simultaneously that Alan thinks the balance shifts and spontaneous rewilding starts to happen.

There are a number of people at the Foundation who have an understanding that we must respectfully ask plants for their permission before we pick them, and I want to know how they feel about Trees for Life employing stalkers to shoot deer. Alan shakes his head. 'Well, I don't know if Trees for Life still do that,' he responds. 'I'm not part of it. Not part of it anymore. I haven't been for almost a year. I worked with a deerstalker. He was a bit more open-minded, but you won't find people at Trees for Life who are tuning into plants the way I was. It's become more of a conventional organisation since my departure.' I want to say I'm sorry but he cuts across before I get the words in the right order. 'When I started it was people from the Foundation, but over time others came, those from a more conventional ecology background.' Sadness pours into his voice and the anger fades. 'I always talked to them. I always said look, this is a spiritual place. We used to meditate. We'd hold hands every morning and we'd call attunements, a few moments of peace and stillness.' He stops to let a lady and her dog walk past and waits until they're out of earshot before he begins again, 'Over time there was resistance and it all fell away. In the end, I was pushed out.'

Ahead of us, on the dunes, little pine trees are growing up out of the sand. It is what Alan wanted to show me. 'Do you ever worry,' I ask, 'that in terms of saving the capercaillie different groups need to be conscious of how they're perceived by others in order to allow us to all work together, whether they're ecologists or gamekeepers or looking at things more spiritually? Is radicalism a good thing?' Alan's mouth softens into a smile and I tell him I saw something about a Foundation member, Franco Santoro, an expert in shamanism, who talks about the sexual experiences he's had while hugging trees, and while he is clearly a man who would support the restoration of capercaillie habitat, his specific passions could be hard . . . – but Alan's point cannot wait. 'I'm a great one for etymology,' he begins. 'Looking at, you know, words. The origin of radical is root. Radical is getting back to the root of an issue. We need that absolutely.'

It is estimated that there are currently around 400,000 red deer in Scotland and Alan believes that for capercaillie habitat to recover in any meaningful way, that number needs to be reduced to some 60,000. In 2004, a group of contract stalkers were helicoptered onto the Glenfeshie estate, deep in old capercaillie country, where they corralled and shot an estimated 80 beasts. The incident is known as the Glenfeshie massacre and was part of a year-long effort that resulted in the killing of over 700 deer. The operation was ordered by the Scottish Government and resulted in fiery protests from gamekeepers and landowners, but despite being vegan since 1979 Alan thinks that the massacre should have been the first of many. 'Even if by some miracle we got our wolves back tomorrow, and there's no reason ecologically that couldn't happen, how many would we get? Fifteen, maybe. That's not going to reduce deer numbers by the amount required.' The sky is darkening and a fug of thin icy rain is blowing in off the sea. We stop where the pines are at their tallest and turn our faces inland away from the stinging cold. Alan believes that the only reason the keepers cared is because they are part of a system that sees deer not as a threat to the pines but as a commodity that people will pay fat fees to stalk. He deeply objects to the suggestion that the deer we slaughter should be eaten. 'You're very righteously saying we cannot waste those carcasses and leave them on the hill, but those carcasses are a fundamental part of the ecosystem. There's a whole suite of scavengers that rely on those. It just shows how disconnected we've become from the whole-ness of life that we see everything as ours.'

In front of us, growing right back to the plantation from where they seeded, pines – 5 feet tall at most – cascade across the earth and run down onto the dunes. Their limbs spread at every angle, every tree different from the one next to it, and no two gaps the same. 'So this,' Alan says, 'is what I said I'd show you. A tiny frag-ment of what we need to see across the country. This is what trees are capable of.' Twenty yards away, an oddity sticks out of the ground, a Sitka spruce, spindly and needles yellow-green. Alan pulls a knife from his pocket and walks over to it. Crouching on his haunches, he hacks away at the base of the young tree with the blade. He works in silence and then stands in an open patch and recites a brief blessing as he throws it across the ground. In his mind, the alien has no place among Scotland's native pines.

Any blue left in the sky is bleeding away and I know that if I want to get there before dark, I'm going to have to go now. Alan hasn't heard of the curse and seems to know nothing of the mythic storm but he tells me the forest is about 10 miles' drive, just across the water if you're a gull but 20 minutes round by road. On the way back to The Field of Dreams, he doesn't say much but I tell him I think I see that saving the few remaining capercaillie will take love from all, and any success will depend on us learning to love each other a bit more too. I worry the words sound hollow but he nods and then tells me about his favourite book in which a telepathic gorilla called Ishmael, hiding out in a San Francisco hotel room, explains to a nameless narrator that in the beginning humans were the 'leavers'. As hunter gatherers, we used just what we needed and then left the land to heal. Then we became 'takers', felling the forests, carving out fields, and saying to any of the animals that if they tried to eat our food, we'd kill them and eat them too. There is hope, though, according to Alan. He's been doing a bit of thinking and he now recognises that whether there's any future for birds like the capercaillie depends on our willingness to progress upwards to the third stage. 'It's as Ishmael says,' Alan tells me as we round a corner and his house comes into view, the chestnut-haired lady watching over us from a window. 'We've been leavers and we've been takers. Now we need to become givers. You need to understand that we can't see every bit of the earth as being ours.'

Little voices echo among the trees, and as I walk towards the track, to set off through Culbin Forest, I see that three small boys are gathered around a young Scots pine. Two of them, screeching, shake the trunk while the third, a birthmark running across his face, hacks away at the bark with a yellow bike pump. As I pass, a man in the car park shouts a string of names and they tear down the hill, whooping with their hands over their mouths like the Indians in the movies.

Labradors, whippets, a handsome apricot poodle and two collies tied to a mobility scooter – the tracks through the forest drown in dogs. Fat old ones plod with a paw at each corner and others tear off under the pines. Half an hour deeper into the trees and the throng of walkers thins, leaving cyclists and joggers. Not really knowing where I am, I turn left up a track. It'll be dark before I hit the water's edge but I want to try and get to where I've read that the village lies buried beneath the sand.

Up ahead, a little girl is zigzagging down the hill, holding her legs out on a bike that may have been the right size two summers ago. Behind her, a lady pedals cautiously. Just before our paths cross, the girl puts her feet down and the mother swerves to avoid hitting her before encouraging her daughter on, but she eyes me strangely as we pass: no bike, no dog, no children.

On my right, red-and-white tape runs across the start of a thin track that cuts into the trees. I look behind me before stepping over it – the lady with the child is glancing back but she turns away before we have to face each other. The track up the slope narrows to a sandy rut and beyond the crest of the hill, 30 yards away, with its base covered in barriers, a large wooden structure rises high above the treetops. Over the entrance there are two large signs tacked together; one reads 'danger, keep out' and beneath it, 'entrance to viewing platform strictly prohibited'. I work out that if I clamber over the handrail on the closed-off walkway I'll probably be able to work my way along, edging round the barrier on the outside before jumping back over to climb the steps.

Once I'm sure nobody is around, I swing my legs across and set off, watching as the drop below me onto the metal struts grows deeper. I move by sliding my right foot along the damp wood before dragging my left after it until the wire fencing juts out in front of me. Feeling my way round, I end up with one hand on the right of the barrier and my other on the left with the metal pushing at my chest. Looking down again, I realise that if one foot slips, I won't be able to hang on, and I grab for the wire with my fingers and pull myself over before tumbling onto the walkway.

The timber beneath my feet is rotting away, which I later learn is the reason the platform is closed. With each stair, my legs feel heavier and I tread slowly, letting the wood gradually take my weight before climbing higher. On the third set, my shoe comes down on something unexpected and I snatch at the rail with both hands before steadying myself and crouching – it's just a bolt beneath a bit of chicken wire and I drop it into my pocket.

Five steps further and I'm on the platform. Behind me, the pines run until they disappear into the gloom and in front they stop half a mile or so away where the sea begins. Somewhere beneath the trees, the first Findhorn Village lies buried beneath a great sand drift. Culbin then was known as the Granary of Moray and the landowner, Alexander Kinnaird, was a wealthy man. But his farms were carved out of an ever-shifting landscape and centuries of villagers felling trees and tearing up marram grass for fuel and for thatching cottages was destabilising the dunes. In 1694, it is said that a great storm blew down from the west and thousands of acres as well as the laird's big house disappeared in a single night. By morning Kinnaird had been rendered a pauper and his tenants were homeless. Many believed that the disaster was the result of a curse. Thirty-two years earlier, Isobel Gowdie, the wife of a cottar, living 3 miles inland, confessed that she'd feasted with the queen of fairies and had communed with the devil who told her that Culbin would be covered in sand. Gowdie was strangled and her body was painted with pitch before being burned at the stake. Four years after the great drift, Kinnaird set sail for Panama in the hope of restoring his riches in 'New Caledonia', but the expedition failed and Kinnaird died at Darien.

Two centuries passed, pines were planted across the vast expanse and not long after that, capercaillie returned. A bird that had been driven to extinction once already across Scotland, clicked, rasped and warbled again in a place that humans had failed to take for their own. They've been gone now since the 1990s, pushed out by little voices, bicycles and dogs, and it's hard to see how any amount of love or fairy magic will ever bring them back. Darkness threatens and overhead, a cormorant drifts across the low sky towards the lights of the streets and houses beyond the cold murk of the bay. I pull the bolt from my pocket and roll it between my finger and thumb before reaching back to throw it as far as I can manage. It makes no sound as it falls into the silence, and then I turn to descend the stairs back into the trees.

The bell tolls for the turtle dove

O yonder doth sit that little turtle dove
He doth sit on yonder high tree
A making a moan for the loss of his love
As I will do for thee my dear
As I will do for thee.

'The Turtle Dove',
an eighteenth-century English folk ballad

Heaped coals glow in the hearth overhung by a knotty oak lintel and rain is being blown across the latticed windowpanes. The start of May has been unseasonably bleak, and tired workers, scraping a living out in the flinty fields, will soon be drifting back into Rusper to drink away the evening. David Penfold sits behind the bar, looking across at St Mary Magdalene, the highest point in West Sussex, where he will one day be buried in the shadow of the sandstone bell tower.

A large figure is tramping down the muddy lane with a box under his arm. Penfold doesn't notice him until he is standing at the window, reading the sign above the door: 'The Plough Inn'. It is the right place. Penfold has never seen the heavy-featured man before and notes, when he asks for a drink, that he is not local. The stranger sips slowly, casting glances at the landlord. 'Do you,' he asks hesitantly, just after he's ordered a second beer, 'do you sing at all?' Penfold smiles. He's heard tell of these young men who wander out, across the country, collecting folk music to take back to town. 'I know a few songs,' he says with a shrug.

The man opens the box and lifts out a wooden case with a fluted brass horn. He explains intricately how the phonograph records sound onto wax cylinders that can later be played back. The demonstration is only a prelude to the question. Penfold looks at the tavern clock above the heavy inn door. It will be at least half an hour until the first drinkers appear. After some thought, he tells the stranger he'll sing 'The Turtle Dove' and walks round to face the phonograph.

From 1903 to the eve of war in 1913, the great composer, Ralph Vaughan Williams, devoted up to 30 days a year to collecting folk songs across 20 English counties. His wanderings were an essential part of the British folk revival which sought to preserve some of the artistic essence of our national heritage. Over a century later, the recording Vaughan Williams made in Rusper on that bleak May day still survives, and though the words are barely perceptible, Penfold's haunting timbre is a rich example of an almost-lost Sussex accent.

The origins of the ballad are unknown but it invokes an ancient conception of the turtle dove as a symbol of everlasting love. Turtle doves mate for life but they love beyond it too, with birds returning to the place of their partner's death again and again, even when their body has rotted and gone. They are also a species that gives the impression of domestic harmony, with the male playing an active role in raising young. 'A Red, Red Rose', written in 1794 by Robert Burns, seems to be a re-imagining of 'The Turtle Dove'. Both ballads feature lovers whose hearts will stay true until the seas run dry, rocks split with the heat of the sun, and stars fall from the heavens. All things that must have seemed unlikely to Burns, writing by candlelight over 200 years ago in a quiet Dumfries backstreet, but feel closer now.

The female turtle dove makes no sound, but starting in spring lone males can be heard purring softly. I don't often think of the call as being a song in itself, so much as I think of it as a rich backing for the sweet whistles of yellowhammers and the trilling of blackbirds, all coming together to form the choir of the English farm. Of the birds we are set to lose, the silence of the turtle dove will probably come soonest. When Vaughan Williams died in 1958, they were present throughout East Anglia and the Home Counties, but since 1995 the population has fallen by 94 per cent. There are now fewer than 5,000 pairs that come back to us across the Channel in spring, and within the next 20 years turtle doves in Britain will purr their last, leaving a great absence in England's summer soundscape.

A version of 'The Turtle Dove', known as '10,000 Miles', has been performed by artists from Joan Baez to the English folk guitarist, Nic Jones. Drawing on the turtle dove always coming back after its great journey down over Spain, across the Strait of Gibraltar, and into West Africa where it winters, the lover in the song promises that wherever they go, they'll return. In truth, though, many turtle doves don't come back because they are shot in their millions. In recent years, the British public has come to associate spring hunting in Malta with the persecution of turtle doves, but any birds heading for England would have to be about 800 nautical miles east of their usual migratory route to fly over that Mediterranean archipelago. But the route back to Africa that our birds do take, as the nights draw in and the leaves start turning, is no safer. In 2021, the French Government finally announced it was going to ban turtle dove hunting after years of refusing to do so, and in Spain, where roughly 800,000 were shot annually, there is currently a 12-month moratorium, but further south, in Morocco, hunters still sit out waiting with their guns. Only some of the turtle doves they shoot will have been born in English hedgerows, but despite not being in a state of decline quite as sorry as our own, the European population is far from well, having plummeted by 78 per cent in the past 40 years.

Plus two-one-two. I punch in the Moroccan dialling code with my index finger and then carefully copy the rest of the number from the top of the website. Beneath a picture of a turtle dove, pink legs perching in the pampas grass and corn, a translated testimonial reads 'strong reactions to this bird that takes to the past'.

Someone picks up. 'Allo! Allo!' I enunciate slowly and talk in an old-fashioned sort of a way. 'Good morning, do you speak English?' The 2,000-mile-away voice is female and sleepy and I wonder if I've got the time difference wrong. 'I speak a leetle.' I picture her in a dressing gown, ashtray at her elbow, sitting by the telephone, surrounded by glass cases of stuffed doves.

'Is this Chassamir, the turtle dove hunting agency?' There is a pause and then she says what I take to be something about me needing to speak to a man called Noel, a cousin maybe. She drops into French and then starts laughing. I laugh too, in a guttural French sort of a way. At which point she asks, in English, if I speak any French, and before I have time to reply she is off, doubtless enunciating slowly and talking in an old-fashioned sort of a way. Then she puts the phone down.

I think about sending an email but the site makes it clear enough. Expect to 'draw 200 cartridges a day' for a bag of up to '50 *tourterelles*'. I have more luck with River Camp Morocco but the man is in a hurry and he doesn't want to hear about Rusper or the English folk tradition, but if I want to hunt, 'sure, send dates'. There are English fixers who I'm told could make it happen, for about the same price as a monthly ticket on the London to Brighton line. In 2019, the *Telegraph* published an investigation into Brits going abroad for the autumn turtle dove shoots. The feature included photos of hunters, guns across their laps and carcasses piled up at their feet, a mass of mottled feathers, cream, purple and blue. There were, as they say, 'strong reactions to this bird that takes to the past', and the fixers went to ground.

We are perched side by side on plastic sun-bleached deckchairs, the blue-eyed farmer and me, in a rusty old shipping container. It is May, not yet mid-morning, but it is the driest May on record and the large metal box will soon be too hot. 'Turtle doves have been part of my life as soon as I could walk,' Graham Denny tells me brightly. 'I'd wander out into the yard and there'd be cattle down both sides, you'd have pigs, and you'd come and there'd be turtle doves sat in the yard. Dad would get handfuls of rolled wheat and rolled barley and he'd throw it on the ground' – he gestures towards the dry earth in front of us – 'and the turtle doves would come and you'd see them there with the sparrows feeding quite merrily.'

He stops talking to blow on his tea and then tells me, hurriedly, that his great grandfather bought Brewery Farm, the 200-acre Mid-Suffolk plot we're sitting on, on the old road to Norwich. We are partly hidden by some canvas webbing that Graham has hung across the front of the container, and about 50 yards out, a few handfuls of cracked seed are scattered across the ground. It was Graham's grandfather who first took him to feed the doves, when he was a little boy, and he's done it ever since. So far, possibly due to storms in the Mediterranean, only a small number of the birds have returned from Africa, but we are hoping one of them might flutter down in front of us to feed before they head up into the trees for the day, in search of shade. Graham tells me he can't remember a time when he didn't love turtle doves, but that the more their numbers have diminished, the stronger that love has grown. It has been exactly a fortnight since the first dove of spring arrived back at Brewery Farm, which was just thirteen days after Graham's father died. 'Heard it purring in the hedge and I just howled and I howled. Turtle doves is something I shared with my dad my whole life and now, when my own boy comes up from his mum's, I make sure he feeds them.'

Graham accepts hunting is a factor in the birds not returning to Britain, but he thinks it's important people respect rural traditions in other countries. 'I'd like to have a conversation with the hunters about it, but it's their thing.' He also feels it's too easy to blame our own failings on happenings abroad. Over the past 50 years, there've been seismic changes in English agriculture, which Graham holds to be the main force behind the decline of the turtle dove. 'The small mixed farm has gone, hasn't it?' he asks me, despondently. 'So many thousands of them gone. If you find a mixed farm, you'll be lucky. We all used to have pigs and we had chickens scratching in the yard.' The small farms he remembers formed a patchwork of abundance across East Anglia, each of them providing a little bit of food and water here and there for hungry and thirsty birds. 'But we aren't allowed to spill anything these days,' Graham continues, 'farming's all gone so clinical and computerised.' I reach for my tea and then, startled by the noise Graham makes, I spill it on my leg. 'Caw-caw!' he makes it again twice and slaps his thigh. A large crow, which had been picking at the grain, takes flight, drifting away over the orchard and out across the meadow. 'Bastard!' Graham shouts after it, in reverence and hatred. He watches the bird until it disappears over a tree line and then tells me that farming's not so generous these days. 'Nature of the beast that big modern farms kill wildlife. You've got a contractor coming in and they turn up with a great 24-metre roll and you've got grey partridges on that field. They're there one day but the next day you go, you're picking up all the dead partridges that have been rolled.'

Graham tells me he is not a wealthy man but if he did things differently, he'd be 'quids in'. His 200-acre holding is split up into 17 small fields 'each with hedges and ditches and bits and pieces' whereas most East Anglian farms are vast swathes of lifeless arable desert, made up of 50-acre fields at least. 'The big boys rip everything up,' he laments, 'and if I hadn't got the love for birds, I wouldn't have any hedges and I'd have about five fields here.' Graham shakes his head when I ask if his grandfather would recognise the farm as it is today. 'It was in his time, when the war came, that the hedges went. All the boys who were farm workers and tended the land went away and the government told them to farm more industrially, but if I took his father for a walk round the farm, my great grandfather, he'd recognise it all right.'

In some counties, since the outbreak of World War II, over half of the hedges have gone. Initially, many were removed, but latterly what was left grew sparse and spindly. Graham tells me, shaking his head dismissively, that restoring hedges isn't rewilding. As he sees it, turtle doves need farming. It's just that they need the right sort of farming. 'We've always balanced nature, for the past 400, 500, 600, 2,000 years. The turtle dove in 1700-and-something was quite rare in this country, so it's only arable that made them popular.' He stops and looks up at the willow tree in front of us. 'Turtle dove, that's a turtle dove in the willow tree, see it?' Up among the leaves, there is a faint flash of grey and I reach for the binoculars.

'Or is it collared? That might be a collared, looks a bit collardy. Bugger, that is a collared.' I place my binoculars on a bag of feed and a bead of sweat runs down my forearm and across the back of my hand.

Graham picks up our teacups and places them on an upturned bucket. By law, a person must check animal traps at least once a day and it is time for him to go on his rounds. I follow him out across the yard and up into the orchard where a magpie is beating its dark iridescent wings against the bars of a cage. The bird is a captive, placed in one half of the trap, in the hope it will lure in others. Graham reaches out and picks an apple. 'You can come up here in the autumn and take plums, pears, apples, anything you might want for a crumble.' The magpie's eyes dart around in fear and Graham glances down at it. 'I had a jay once,' he tells me, as it hops over to the other side of the trap and drinks out of a small water container made from a plastic milk carton. 'It could make a noise like a buzzard. I think he learned it so he could keep other birds out of his wood, but truthfully you've got to take the rough with the smooth. Pest control is a sort of necessary nasty. Corvids and squirrels, they eat the turtle dove eggs. They'll take chicks.' He looks at me then throws his arms up as though to say, 'What else can I do?'

We walk slowly down to the barn, where a rusty farm pick-up is parked out of the sun. On our left, over the fence, an owl is hunting mice in the meadow and Graham watches it while telling me about the cows they used to have. No conversation about Suffolk goes by without someone who holidayed there once remarking on how flat it is, but it's not entirely true. We wind our way along a road that borders Brewery Farm and the land slopes upwards on either side. Where we turn off, Graham points to a field on our right and tells me that he planted it yesterday morning with a wildflower mix. In the shadow of a sprawling hedge, he brings the truck to a stop and knocks it into first gear to hold it on the slope. The pullout hand-brake looks as though it stopped its last some years ago.

'What you want to do,' Graham tells me, looking to a gap in the scrub, 'is stand on top and you'll be able to see it all.' I push through the thorns and scramble up a mound of spoil. 'You see,' he shouts, 'ash, hazel, bramble, spindleberry, dog rose, old man's beard, buckthorn, blackthorn!' It's as though he's a druid reciting an incantation. Stretching out in front of me as far as I can see is a great tangled thicket. In centuries past, hedges just like it would have marched across England providing endless nesting habitats for turtle doves. 'It's about 15 ft wide!' I shout down. 'That's as much as 20 in some places!' he shouts back up.

I jump off the spoil and we walk along the hedge before cutting through a gap into a hollow. 'I chucked some loppings in there the other day, see, at that turtle dove nest site.' Graham thrusts his unshaven chin forwards to a hole that would go unnoticed to an unfamiliar eye. 'Secretive birds,' he says, in a half whisper, 'you only hear them when they want to be heard. When you get a male calling, he's looking for a shag.'

Back in the truck, we thud our way across the ruts and then grind along a stony track before pulling up at one of the farm's feed stations. Whenever he has a spare couple of hours, Graham drives round neighbouring farms to pick up any rapeseed husks for the turtle doves. The first station he built was in 1996 and all of them are covered with corrugated-iron sheets so the birds can feed happily without the threat of aerial attacks from sparrowhawks. In decades past, weed seeds in unkempt corners kept farmland birds fed, but in a world where almost every inch of available ground is turned over to crop production, starvation looms. While I'm rooting around in the rapeseed and letting the husks run through my fingers,

Graham tells me that half the trouble is 'we're just too bloody tidy nowadays. Things like bramble, that's a hell of a habitat, but people won't leave it be.' I follow as he walks towards a thicket and points into the dense prickly mass. 'Little yellow bills, see?' Among the dark thorns, I can just make out three long-tailed tits poking their heads out of a nest. Graham tells me they'll be fledging soon, and then we carry on across the fields on foot.

In England, subsidies paid for agricultural land include hedges but it is stipulated they can be no bigger than two yards wide, beyond which they are categorised as scrub. Until 1998, Graham remembers people being paid for scrub removal. 'It was madness,' he continues, 'they were actually getting money for habitat destruction.' He points to a hedge in front of us. 'I lose money on that, £250 each side for encroachment. The guys in the Government office, they don't understand the countryside whatsoever. They are some of the worst bird killers there is and they wouldn't know what a bird was half the time.' The trouble, as Graham sees it, is that they want to put the countryside into little boxes, 'they say you get money for grass and money for hedge but scrub is neither'.

We cross a slow-flowing tributary of the River Orwell and Graham softens. Running up a field margin in front of us is a hedge three times his height and he stops to brush his hand through it. 'It's sort of rewilding itself,' he turns and says to me. The word brings me up short and I'm about to say I thought he didn't go in for rewilding, when a leveret breaks from a form beneath our feet and pulls away hard across the plough. 'Chivy,' Graham says with a smile, as he watches it go. 'What's that?' I ask. 'That's just leveret, Patrick, in Suffolk,' he replies. 'I like to remember those words.' On top of the hill there is a small now-doorless building where Graham went to school when he was a boy. He tells me he could look out of the classroom windows and see everything that was happening on the farm. In those days, it supported his father, his grandfather, and his two uncles. It wasn't long ago but Graham tells me it was sort of a different country then.

Back in the yard, side by side on wooden stools, we drink more tea in the sun. Year after year, turtle doves will return to the very same nest and Graham thinks it means that you get to know the birds. As he talks he punctuates his thoughts by tapping a spent rifle case on the table and I wonder about the animal that met its end when the bullet lodged itself in its brain or blew its lungs out. 'It's like the swallow coming back to its barn or the swift going back to its nest. Just by where we're sitting three years ago,' he points at a hedge. 'A bird ringer rang two pullets – female turtle dove chicks – and the following year, someone saw them down at Welney on the Fens and was able to identify the number on the ring with a telescope. The following week, the birds turned up back here. They even wrote about it in the *East Anglian*.'

Up a ladder, by the door to Graham's cottage, a young man is running a paint roller across a once-white wall, a t-shirt tied around his head to keep his long hair from his eyes. 'The turtle dove is all part of it,' Graham says, looking round to watch the boy on the ladder. 'It's all part of the countryside. It's a familiar sound of spring.' He turns back to me and runs the back of his hand across his cheek, wiping away the sweat. 'I don't just see green trees and green fields. I see everything here and I love everything here. If I see a rabbit,' he says, glancing down at the case, 'I might shoot the rabbit because I want to eat it. I don't hate that rabbit. I love that rabbit.'

Graham has drilling to be getting on with and I tell him I'll just run over to the shipping container to get my binoculars before I head off. 'Walk slow,' he replies, 'there's sometimes a turtle dove feeding on the grain.' I creep along the side of the barn and peer carefully round the plastic downpipe. There is a flash of russet and gold but it is only a yellowhammer and it flits away over the hedge. But then, in the orchard, I hear the soft rhythmic purr. It rolls on and on, and I think of Graham as a man fighting a losing battle to the end. There will be a time, maybe even before his own boy grows up, when he is standing in the yard, one June morning, and it suddenly becomes clear that the turtle doves are not coming back. In that moment, England will lose part of its soul and Graham will lose part of himself.

In the back of the shop, a large man leans over half a pig, pulling a knife downwards across its fat belly. 'Best butchers in Suffolk,' Richard Negus announces, as though he has been sent by local radio to bestow a grand prize upon them. He pulls open the door, a bell sounds, and the man turns while pushing his glasses up his nose, the blade passing his cheek. Another even larger man, standing behind the counter, takes us in with a wide smile. Richard gestures towards the chopping block. 'Jamie there, poor sod, he ended up butchering a big red hind I brought in.' Jamie nods. 'It was the best sixty quid I think I ever spent,' Richard continues. 'Worst bit of business we've done,' the fat butcher replies, turning to grin at me. 'We didn't see him for months.' Richard is stooped over, addressing the sausage rolls sweating in the bottom of the cabinet. 'Finished the last bit of venison only last week. We were eating it every which way.'

Jamie is pushing his glasses up his nose again, eyeing the pig's kidney, and the big butcher is chortling away, fingers clasped on his paunch. 'Shall we just get two of those, Patrick? I'll get them.' I wander round the shop, looking for a can of coke or a machine that splurts cheap coffee into polystyrene cups, but they only have garden mint and pomegranate cordial or apple juice from Somerset, 'organic, cloudy and crisp'. We are taking too long and an old lady who slipped into the shop behind us has run out of patience. 'I want two chops, four sausages, and half a pound of bacon.' Her face is sour and the Saturday boy, lingering by the back door, lurches into action.

Outside, on the pavement, a little girl is lifting a fox terrier's legs up off the ground as though the two are dancing. The dog looks forlornly at the hardware shop, waiting for its master to return. Richard and I walk down the high street, greasy flakes of pastry falling down our fronts as we eat. 'Do you ever see *Lovejoy*?' My mouth is full of porky ballast and I shake my head. 'It was filmed here. It was one of those gentle, Sunday night, comedy dramas.' When Richard speaks he closes his eyes, leaving you grasping for some sort of greater meaning. 'Lovejoy, he was just on the right side of bad and he had a mate called Tinker.'

Sausage roll in his left hand and right hand on the wheel, Richard heaves his old truck round like he's out in the Wash turning a boat against the tide. On our way out of town we pass a row of timber-framed houses before dropping down towards Westhorpe. He is talking about *Lovejoy* again when the phone rings. 'I have parcel but can't find your house.' The man sounds young and annoyed. 'Whereabouts are you sat, my friend? What can you see?' There is a moment's silence, presumably as the delivery driver looks for any distinctive landmarks. 'I have on my left Church Green.' Richard nods. 'Is there a silver pick-up?' The voice on the line cuts back, angrily, 'No, I have just delivery.' In confusion Richard shakes his head and I wonder whether I should join the conversation when the voice sounds again, 'I find.' Richard smiles, 'That's it. You're a good old boy,' but the man has already hung up.

Five minutes on, we pull up in a gateway. Out in front of us, a kite is hunting low over the plough and I get out and watch it casting circles while Richard rattles through the boot of his truck. He leaves it open and comes back round, holding his billhook, a dark curved blade glinting silver at the edge where it's been sharpened away. As he crosses the field, towards the boundary hedge, he looks up at the bird and murmurs under his breath. He walks at about twice the pace I do and I shamble after him across the mud. 'Here in Suffolk,' he tells me, 'when I first started laying hedges about seven years ago, farmers would go, "Fuck off, hedges are no good to man nor beast." We don't need that.' When Richard was young he joined the army as a soldier but was soon encouraged to apply to Sandhurst for officer training; a love of horses eventually led to him commissioning into the Household Cavalry. His speech meanders from the rich vowels of his boyhood in Southwold – where his uncle Gary is now the last trawlerman at the Blackshore Quay – to the cut consonants of the officers' mess. 'Something's changed though,' Richard continues, bending down to grab a twist of blackthorn. 'There are lots of factors and I can't quite put my finger on it. One thing is that in the past few years a spectrum of people have truly started realising how important hedges are for conservation.' He runs his gloved left hand over the bark, then bends the shrub down to the ground. His right arm moves back and forth with the blade as though he is finding the perfect place, then in one motion he brings the billhook down on the blackthorn, splicing hard in the direction of the grain. 'Largely, it's that they're sniffing out grant money from schemes as opposed to becoming environmentalists' – he stays crouched in the hedge but turns towards me. 'Round here, people sneer. They don't want to be seen as hippies. They don't like all that shit. They don't want to be seen as a Barker.' He nods his head in the direction of the farmhouse beyond the lane.

'Mabel!' Richard shouts. His little black dog has scrambled out of the truck and is hunting in the grass. 'Come here!' She snuffles towards us and he continues, 'The current environmental schemes are revoltingly complicated. There is money to be had but you've just got farmers going "Christ's sake". They can't crack it. The ones that are really canny though have listened to what they've said about ELMS, the Environmental Land Management Scheme, and realised that an agricultural revolution is about to hit Britain.' When he opens his eyes again he looks at me and nods, as though he's about to tell me something that will change the way I see the world. 'What it is, is that now we've left the EU, agricultural subsidies are going to change. Rather than paying lip service to a bunch of conservation measures, there is a number of pillars you'll be judged on.' He slows and counts them out on his fingers. 'The environment, carbon capture, water quality, heritage and history,' then he looks blank. 'What's the other one?' I shake my head. 'It's some other bullshit but it might be for example that you have an ancient woodland and you're going to manage it because it's, well, because it's ancient, so you keep that in good nick and you get your money.'

He beckons me over and I sit down next to him on the warm sandy earth. It is early autumn and the leaves on the blackthorn are starting to fade to yellow. Across England, the last of the turtle doves are taking flight for Africa, and the hedge-laying season, part of the restoration of their lost habitat, has only just begun. 'That, now, is a pleacher,' Richard begins. He draws out the strange word. 'It's a sort of hinge, a cut, that allows the sap to continue flowing.' He bends it into the hedge. 'That's your first pleacher laid, so that'll grow the whole way along and the blackthorn will carry on growing out of the heel so the whole hedge thickens up. We'll do another.' He repeats the process. 'So this one goes over the top of the next, and then the next one goes over the top of that. Then when I've

189

got a certain amount done, I put a hazel stake in every metre or so and that retains it.' He speaks rhythmically as he works but then stops and looks up at me flatly, as though trying to work out if I'm following at all. 'Final thing is binding. I bind it along that line with twine and I twist it over and over and over until it holds it in a rigid form. The whole hedge becomes much thicker. Twenty years that'll last, when it's all done.'

Richard stands, puts his hands at the base of his spine and stretches with a pained groan. 'Hedge-laying is easy, really. The hard bit is getting any good at it and it's back-breaking.' He pulls his sleeves up to show me his arms. 'They're a scratched-up mess and they never heal.' I look at the angry skin, thinking that his body is breaking down as he breathes life into the landscape. He shouts again at the dog and she spins round and follows guiltily on as he sets off along the hedge. Richard admits that you could leave it. 'It might grow up into reasonable turtle dove nesting habitat but my argument would be, if you give it a firm foundation by laying it, it's going to have more strength to be able to grow up to the height they want it. I'm not saying that because it's my job. It just makes sense.' He smiles and I wonder how many times he's said those words while pitching for work. Richard points to the blackthorn with his billhook. 'Those gaps there, you can peer straight through.' I look down to see Mabel bounding along in the field next to us. 'And I tell you what, that's about 98 per cent of hedges in this country look like that.'

Fair-haired Patrick Barker is standing in the yard, binoculars hanging round his neck and a puffin embroidered on his fleece. In the 1990s, he would have been a catch at the Suffolk young farmers' club. I wander across the concrete and he almost manages a smile. 'So I've been hedging for Patrick and Brian for almost three years now, is it?' Richard says, by way of a hello. Patrick nods. 'Is Brian your brother?' I ask. 'My cousin,' he replies, walking towards his pick-up. 'It's two generations farming together, two mums, two dads, and two of us, the boys.' He jumps in, and I get in across from him. 'I'll maybe leave my truck here,' Richard suggests before clambering over the feed bags onto the back seats. The big diesel engine chokes to life reluctantly and we set off on a track through the middle of the 1,200-acre farm. Patrick drives in silence, looking out over the fields, with his elbow resting on the open window. Up ahead of us, in a gateway, a covey of grey partridges takes to the wing, their 'kut, kut, kutting' call just audible as they flutter away before drifting out of sight, tight against the contours of the barley stubble. His gaze lingers for a moment on the point where the partridges disappeared and then he turns back and says that his grandfather bought the farm in 1957. 'He lived at Hall Farm, Langham, about 8 miles from here. His neighbours, the Blackwell family, as in the books, made him an offer on the farm he couldn't really say no to.' We park up beside a row of crack willows on the edge of a pond. Patrick stands for a while, looking down at the dry bank where the water has dropped away in the heat and then tells me he thinks there was a lot of moving around post-war. 'Lots of young men had disappeared and the older farmers stayed working. When they died, there weren't many people to take over so the land was sold.'

Richard has dropped off and is 100 yards ahead, looking along a thick 5-foot-tall mass of woven green. 'This is one of mine, done about March time,' he shouts back to us. We turn and walk towards the newly laid hedge. I tell Patrick I was impressed by how much Graham Denny had achieved at Brewery Farm and was pleased to hear he had four pairs of turtle doves in the end this summer past.

He looks up from the earth in front of us and smiles thinly. 'Over the past fifteen years, we've gone from having four or five pairs here, across a thousand acres, to maybe having one.' Patrick pre-empts my questions and tells me no, they couldn't be doing much more. 'We're doing lots of things, all the environmental work and the number of species we're looking at. All the things we're trying to balance on a day-to-day basis, farming and running a business. We're supporting three families. Graham's farm is a bit of a nature reserve. He drives a digger on the side to earn money.'

When we get to Richard, he is stooped over, running his hand through the hedge. The money a hedge-layer can make, he tells me, is reasonable. He calculates as he talks. 'Thirty-five metres flat out. That's by myself. It's £9.40 a metre, laid and staked, and if I bind the top it ends up at £12.80 a metre. Traditionally you do "a chain" in a day, that's 22 yards, so really 25 metres a day would be normal. That's fifty in two days and on you go.' He pauses and casts his gaze along the hedge as though he's admiring it but his eyes are closed. 'I'm really quite pleased with this. Wouldn't you say it's smart?' On opening them he breaks into a wide smile and I turn to follow his line of sight. Unable to scent us, with the wind at their backs, a roe doe and her fawn are gambolling across the stubble. When they are 30 yards out, the mother stops and her body stiffens. Richard gives a shrill whistle and they turn, the doe running on ahead towards the trees and the fawn tearing after her.

Beneath our feet, as we head along the top of the field, is a thirsty-looking 6-metre wildflower margin. 'The plan is that it will attract beneficial predatory insects,' Patrick explains. 'We want to look at how they affect pest insects on the crops in the hope that farmers will be able to create areas that house and capture those insects as opposed to being reliant on sprays.' Richard and Patrick walk next to each other, staring at the earth around their boots as though looking for something lost. 'It's not designed with turtle doves in mind,' Patrick continues, 'but the weed seeds in here should be ideal for them too.' He stops and pushes his hands into the pockets of his fleece. 'All of this would have been seven or eight fields once. Every field would have only produced about a tonne of wheat an acre. The rest would be poppies and marigolds, things that would have fed the turtle doves through the summer, but we've become too good at crop production. We're too effective.'

At the top of the headland the ground starts to slope away. 'Chickweed and milkweed,' Richard announces. 'They need things like that, things that are seeding as soon as they arrive in spring.' He crouches and picks some leaves from beside his boot, a thin green stalk and pretty white flowers. 'Shepherd's purse' – he says it as though it is the title of a long story he's about to start and Patrick cuts across him, 'They're vulnerable, chickweed and milkweed, one dose of spray and they're gone.' We pass another pond and the farmer and hedge-layer continue looking down, as though concerned that the other will beat them to identifying something of significance. 'What's that stuff?' I ask. 'Fumitory. That's meant to be the turtle doves' favourite, isn't it?' Patrick shrugs. 'To be honest, all they did was had a load of turtle doves in a cage and they put down bowls of millet, rape and fumitory, and the turtle doves wandered over to the fumitory.' Richard makes a knowing noise. 'Nobody actually really knows,' Patrick continues. 'The chap who runs Operation Turtle Dove who carried out that test, he's a friend. He said to me ten years ago, if you can actually work out what turtle doves really need, you'll be the saviour of them.'

Dense and scrubby, rising on our left and running towards the village, is a row of thorns. 'Now this,' Richard tells me, 'is really very interesting. The thing with laying hedges traditionally is that it can be hugely expensive. That hedge there,' he points back in the direction of the one he laid in March, 'that's £2,400, a big old lump. Prohibitively expensive for most farmers if they don't have a grant. But for this, which is dead-hedging, you only need to hire a tree shear and I charge just £200 labour for the day.' Somewhere on the other side of the tangled mass a creature bolts, and as it bounds away, its feet make a dull thud against the earth. 'Now this tree shear' – it is his moment and he speaks as though he's standing up at the front in a lecture theatre – 'it's like a giant pair of secateurs. What it does is it cuts the blackthorn, which then grows back with lots of shoots. Then, with its sort of mechanical claw, it places the brash back over the top to protect it from browsing by deer and hares. We could do this whole hedge in a day.' I squat down and reach through, trying to make sense of it all with my hands. Richard hovers above me, making an approving noise in the back of his throat, and Patrick stands 10 yards away, picking at his nails. 'It was a traditional Suffolk method,' Richard tells me. 'When a hedge got completely out of control, they'd quite often coppice it down to the ground.' Grasping around in the dead thorns, I find tangled green growth sprouting. 'It fell out of fashion with the invention of the flail mower in the 1970s, but its history is ancient. If you look at Caesar's first invasion, he writes about Britons using dead-hedging to stop legions of Romans. At Maiden Castle in Dorset, each ridge had a line of it. You just think, if you had the old legionary kit on and you're trying to scramble over this but it's three times the height . . .' In my mind's eye, I see those Roman soldiers from the Asterix books I kept under my bed as a child, large noses and leather sandals, drowning in great hedges, while the Britons bear down on them with axes. I tell Richard I like the idea of England using dead-hedging to protect itself from a conquering force and him now using it to try and save a species that is integral to what Britain is. Patrick looks up, makes a faint noise, then goes back to his nails.

Richard walks down the hedgeline to the bottom of the field, talking admiringly, half to me, half to himself, about the rate of regrowth, and Patrick and I wander along next to each other in silence. A half a mile away, on the other side of the village, a tractor is driving up and down, perfect lines on a perfectly rectangular field. Patrick tells me that over the years most of the hedges on the farm were kept low so they could be trimmed with a hedge cutter. 'A lot of the reason why turtle doves struggle to find nesting habitats in this country is because the height of hedges is dictated by the machinery used to cut them. If a hedge is too high to manage with a flail attached to a tractor, it's basically a decent turtle dove hedge. They want height.' It has been 15 years since the current generation of Barkers started restoring the farm's lost hedges. 'There was an old boy working for us, well into his eighties,' Patrick recalls, 'who just laughed when we told him our plans. He'd been around to see our grandfather bulldoze them out in the 1960s to enlarge the fields and make the land more profitable.'

I want to know why it all matters to Patrick so much, all of the habitats and everything that lives in them. He bites at his lip and shrugs. 'I have an interest in wildlife. We just want a farm that's full of wildlife.' Richard is listening in and laughs when I ask Patrick if he finds that other farmers feel the same way. 'I went to give a talk on hedge-laying at the Suffolk Machinery Club,' Richard replies. 'I half think they just wanted to find out what was happening here through me and they did genuinely say, you won't talk about the Barkers too much, will you?' Patrick smiles joylessly. 'Funny old boys,' Richard continues. 'It's a club for those who like big tractors and big machinery, any development that can help you eke out a few extra quid.'

I fall behind to pick a handful of brambles. 'I was there the other day,' I overhear Richard saying to Patrick, while they are crossing a stream in front of me with willows running along either side. 'One field had almost fourteen corners and everywhere we went he just kept on saying, remember that this is a commercial farm. Just remember, we're commercial. It was just utterly bizarre.' When I catch up, the conversation ceases. 'So this, historically,' Patrick begins, 'would have been a meadow. It was farmed from the 1950s until about 2007 when we took it out of production. It's surrounded by hedge, we've got a pond right in the middle, there's a footpath and there's houses.' A hundred yards away, across the knee-high grass, a lady stands in her garden, pegging clothes on a washing line. 'The land,' Patrick continues, 'in terms of costs, isn't worth farming.' Richard ducks under a willow tree, grabbing a branch so that it doesn't fly back towards us. 'It just all depends on what people think of as productivity,' Patrick tells me. 'If using every last bit of land is thought of as being productive, then there's no hope for turtle doves.' Three cabbage white butterflies flutter up out of the grass in front of us and blow away over the meadow. 'If we farmed this land,' Patrick continues, 'we'd have more wheat in the shed this year, but because of the difficulties that the stream, the pond and the houses create, in terms of using machinery, it would cost us twice as much to farm this as it costs to farm that.' He points towards a large field back the way we came. 'To my mind,' Richard adds, 'farmers should farm to the best of their abilities, but where that isn't a sensible thing to do, they should make a place for wildlife.'

We emerge from under the trees into a clearing where piles of seed are spread out across the earth. There are six areas in the meadow where they keep feed out all summer, just in case the turtle doves do come back. Around us the hedges rise higher than the telephone poles running across the fields. We cut back by the garden where the clothes are flapping on the line, and a cat, at the window, watches as we pass. Patrick tells me there was only one turtle dove that came this year and even then nobody saw it. 'It was only the sound.'

A group of middle-aged men in lycra are pedalling hard into the wind and Richard waits for them to pass before pulling out of the farm gate. Mabel, on the back seat, whines softly. I want to know if Richard thinks the battle is lost. He turns to the dog, telling her to be quiet. 'I think it's about choosing the right battles. Look at everything they're doing there,' he glances in the rearview mirror. 'This year, the sum-total of all their efforts was one possible turtle dove.' Two fields over, a kestrel hovers, hunting above a grassy corner, and I watch, hoping to see it stoop, but we round a bend and it disappears from view. 'People think they know what the turtle dove needs,' he adds, 'but they never say that what it really needs in England is for a third of the population to be culled off and all the roads and towns and shopping centres to go with them.' For some minutes, we drive on in silence and then we cut left into the houses and I realise, on reading the street name, that we're back at Richard's cottage. As we drive up Church Green the sun is at its highest point in the sky and children are playing on the grass. 'It would be remiss to do nothing for the turtle dove,' Richard says, bringing his truck to a stop, 'but if we were going to divert dwindling funds into the difficult fix of trying to have them and in doing so we ignored other less beautiful species like the tree sparrow, I think it would be equally remiss.'

Just before he gets out, Richard turns to me and smiles knowingly. 'It does feel as though there's no soul there with those boys, but there's so much passion beneath it all. Farming is a masculine thing. It's isolating, high suicide rates. But I think what it is, the reason they do it, it's the quiet moments of wonder. It's the silent moments of awe.'

Afternoon is giving way to evening and autumn is becoming winter. Over the road, high in a sparse fir tree, silhouetted against a dark racing sky, a crow clings to a branch, calling on the wind. It is too cold to drink outside, and as I push at the door, swollen wood scrapes against the flagstones.

When I'm lost at sea I hear your voice
And it carries me.
In this world we're just beginning
To understand the miracle of living.

Belinda Carlisle, 'Heaven is a Place on Earth', number one for two weeks in 1987. The sound system is fixed to the wall above the knotty oak lintel, but there are no coals glowing in the hearth. The landlady sits behind the bar, talking to the only other drinkers. 'It's the best thing as I say, but as he says' – a woman, mid-fifties, raising her voice to be heard above the music, nods towards the man standing next to her – 'it's hardly going to be the same.' The man puckers his mouth like a dog's arsehole and shakes his head despondently, while gathering the empty glasses from the table. 'You're not going to leave me, are you?' the landlady asks. 'Afraid so,' he replies. 'We've got to be off in the morning.' Through the window, I watch as they wander away up the lane before disappearing at the crossroads.

I sit, finishing my beer, looking round the room while avoiding the landlady's glances just as she is avoiding mine. If I don't go now, it will be too dark. Over the road, down a lane, left along a narrow path covered with the last of the acorns, and through a gate hanging off its hinges. At the first gravestone, I can just make out 'IN LOVING MEMORY 1798' but the rest is covered in ivy.

Beyond the shelter of the church, the wind hits me, blowing straight through my canvas coat. Looking out across the graveyard it occurs to me that, at dusk, stone crosses look like square-shouldered men. I turn and look back. Halfway up the tower, a shattered window has fallen open, and next to it a rotting bird box is nailed to the wall. Casting my eyes around my feet again, I realise I'm in the wrong place. All dead far too long ago, names mostly faded on the crooked headstones.

I walk further, among the trees, until the writing becomes visible: 'RICHARD KING WAS LOVED DEARLY, JAMES PECK LOVED HIS WIFE MARY ANN', and 'EMILY GOWER FELL ASLEEP IN 1932'. None of them are what I'm looking for. On my left, carved out of a bit of field, is the newest part of the graveyard, death taking over. There are only about 20 stones, each one of them different, and I stop in front of them all. Towards the end of the second row, my foot hits a grave lying flat and I stumble: 'LUCY BROADWOOD, DIED 1929'. Next to it there's a cherub, hands clasped at the side of his sleeping head and plump little arms green with moss. Beneath, on a plinth, the writing is just visible: 'ERIC VALENTINE, DEARLY LOVED YOUNGEST SON, MARCH 1932'. As I read, I can see, further down, almost hidden, there is faded lead lettering, some of it missing. I pull the grass aside and run my fingers across it, reading the words and tracing the gaps – 'AGED 4 YEARS: IN HEAVEN'S SUNLIGHT SAW NO SHADE OF FEAR'.

Somewhere in the village a door slams, and I don't feel like walking back among the graves so I head down to the fence to try and find a path running round the outside. At the bottom, I get to a hedge and I walk beside it in the direction of the lane and the lights. Carried on the wind, I can hear the sound of cooing wood pigeons roosting out of the cold. As the lane comes into view I quicken my pace before noticing, on my left, through the bramble, a gap that opens out in front of row upon row of more graves. I climb through and squint in the half light. I had made peace with not finding it but the name I wanted is carved on a small granite slab in front of me: 'PENFOLD, DIED 27 JUNE 1939'. On moving closer, I realise it's not right. It reads John Penfold, not David. I look out across the headstones, wondering whether the man who stood across the road in the Plough Inn, 113 years ago, singing a ballad inspired by a bird that no longer comes to Rusper, really is lying beneath my feet. Maybe he had two names or maybe Vaughan Williams got it wrong. For a moment, I think about going from grave to grave, but it's too dark and the bell in the tower begins to toll.

Another church

Tumbled in bag with rabbits, pigeons, hares,
The crumpled corpses have forgotten all
The Covey's joys of strong or gliding flight.
But when the planet lamps the coming night,
The few survivors seek those friends of theirs;
The twilight hears and darkness hears them call.

John Masefield, 'Partridges', 1936

Beneath the saintly toe, to the right of the dunnock, the grey partridge stands, glass beak pecking at the leaden ground. I walk three steps back into the cold quiet to take in the whole. Above Francis's face, thin lips and skin sallow, swallows and swifts cut through the air, and down among the folds of his robe finches flutter, gold, green and haw, their heads turned upward in adoration. On his forefinger a robin rests, and out on the left a barn owl perches, wings outstretched like an angel. Every bird in the window, installed exactly 100 years ago, is taken from the pages of Gilbert White's 1789 book, *The Natural History and Antiquities of Selborne*, the final chapter of a life spent in thoughtful observation in a thatched-roof barrel.

I look round at the altar where White's father and grandfather preached, at a time when men believed house martins spent the winter hibernating in muddy pond bottoms, but a noise beyond draws my attention back. Out in the churchyard, above the lichen-pocked stone marking Gilbert White's body, the rooks are rioting, drunk on the warmth of the high summer sun. Farther still, where the cottage gardens end and common grazing begins, White often saw partridges in 'vast plenty', in the shadow of the still-standing beech wood. So prolific were they, that in 1740 and 1741 'parties of unreasonable sportsmen', much to the naturalist's dismay, 'killed twenty and sometimes thirty pairs in a day'. On the edge of the trees, where White's wooden hide stood, a re-creation stands now. You could sit in it all summer long, but there are no grey partridges left to see in Selborne.

Three steps forward and I run my finger along the bird's brown back and down across its plump grey neck, speckled blue. Its glass feathers glimmer in the early evening light, but its eye is pale and forlorn. At the other end of the church, a priest loiters in the sand-stone doorway, waiting to lock up behind me. 'You see that,' I ask, pointing up at a metal chain hanging from the ceiling, trying for more time, 'is that for incense?' The priest stays where they are. 'There might have been a thurible on the end once, but we don't use it now.'

In the late 1970s, in a Basel suburb, a six-year-old boy appeared at the bottom of the stairs, midway through his parents' dinner. He had been thinking on it and he wanted to let them know he'd decided he was going to become an ornithologist.

The heat is rising over newly cut wheat stubble and the kites are riding the thermals, hunting mice whose world has just changed.

We are 7 miles west of Selborne, on the western reaches of the Sussex Downs, looking out across the rolling Rotherfield estate. 'That is what my parents tell me anyway,' Francis Buner says, with a shrug, before he breaks into a dubious laugh and walks off down the field, beneath the telephone wires. 'There was so many of them', he continues – all happiness in his voice gone – 'that you have no imagination how many millions there were of grey partridges in Britain. It was the most common farmland bird.' In a paddock beyond a cottage garden, to our right, a horse whinnies and Francis turns in its direction, as though affronted by the animal's interruption.

Little Francis hadn't been wasting his parents' time. After studying music he went on to do a degree in zoology, before narrowing his focus for a Masters on kestrels and vole abundance, but it wasn't where his heart truly lay. Towards the end of his time as an undergraduate, he had been to see his tutor to tell him he'd come across a picture of a willow grouse in a book he'd happened upon in the library and wanted to make a contribution to the study of the species. He left the meeting having been convinced that the willow grouse wasn't the one but that the grey partridge, which was fast heading for extinction in Switzerland, was a bird that needed to be fought for.

There is no breeze and the sun is at its highest in the endless blue sky. As we wander across the dry earth, Francis's thinning curls start to darken with sweat. 'Of couse, I wanted then to do my Masters on grey partridge, but in the end it was simply too early in the timescale, but I did see one.' He speaks with a matter-of-factness, reeling at speed through thoughts and recollections that run out, every so often, into moments of deep joy. 'For my Masters, eventually I focused on kestrels and I had to be trained in radio tracking for which I was based in Lincolnshire. We did at first a pilot radio-tracking study on turtle doves. These turtle doves, they are absolutely wonderful. There, in Lincolnshire, on that farm, I saw my first greys.' Above us, rising and falling across the fields in the thick air, a wood pigeon flutters upwards then stoops. 'When I got my eye into it,' Francis continues, 'I saw greys across the border from Basel in Alsace, where they still have some, but the next ones I saw was in the study area for my PhD.' His eyes widen and he starts laughing wildly. 'I started from a very small basis, no?'

We stop at a sign beside a telephone pole, where foxgloves are growing up among hawthorns. Two bees fly round and round the purple trumpet flowers as though making sure they have left none unvisited. 'It's funny so many people call them English partridges,' I say to Francis, 'when they're spread all over Europe.' He smiles. 'It's very English this, but they are only called the English partridge because the red leg partridge is called the French partridge and this French partridge, which is in Italy and Spain, is only called the French partridge because King Charles II, unfortunately for him, had to come in the 1600s-and-something, I think, and rule England. He loved to hunt the red legs and he bring the red legs from France and that's how they came.' To the left of the sign, a sculpted grey partridge, its beak open as though it's calling across the fields, is nailed to the top of a post, rusting in the sun. Game records running back to the 1840s show that the Scott family, who still own Rotherfield, used to shoot between 10 and 20 grey partridges, three or four times a season, but by the 1990s, largely due to agricultural intensification, the birds had disappeared completely.

Francis looks down at the sign. He walks the fields most days and he wrote the words, but he pores over them with the fascination of someone who has never before ducked beneath a gate. 'It's quite fancy, I think it's fair to say. This sign is to show passers-by that the people running this project really care. It's not just some sort of . . .' He breaks off, trying to find the right word, and then goes on a different tack. 'There is ambition here. This is serious.' Francis moves aside and steps back, carefully tucking his t-shirt into his jeans while beckoning me forwards to read about his work. On the left-hand side, black letters in a blue bubble read, 'Since the 1960s, the grey partridge has decreased by more than 90 per cent. The main cause of decline is the loss of nesting habitat and the use of chemicals to combat weeds and insects on farmland.' Beneath the words, two little rows of 10 partridges, grey silhouettes, march into the middle of the sign with the caption, 'Britain, 1900: 1 million pairs'. Ahead of them, beside a green butterfly, a lone bird walks, with the words, '2009: 36,000 pairs', written in red. In the middle of it all, watched over by a buzzard, a partridge in fine detail, wings set on an elegant diagonal, flies above a harvest mouse.

In 2004, following a study confirming that Rotherfield had no remaining greys, the estate commenced an innovative reintroduction programme in partnership with the Game and Wildlife Conservation Trust, where Francis is a Senior Research Scientist. The plan was to release pairs of truly wild adults, which would be placed with a brood of reared partridge chicks. The hope was that, in time, the adults would foster the young greys and teach them the necessary skills to survive in the wild before they themselves would go on to breed the following season, eventually re-establishing a self-sustaining population. Progress was slow, but after a decade of releasing and fostering, Francis was regularly counting over 50 chicks which were being born naturally across the estate. I turn around to see that he is pulling his t-shirt out of his jeans. It seems he has decided, after all, that it would be better off not tucked in. He walks towards me and places his hand on top of the sign before spinning it round, with a look of wonder on his face, as though it's magic and he has just pulled a rabbit from a hat.

We set off down the field and Francis explains that hot summers and wet winters, which have lately become more frequent, result in chicks dying of thirst and cold. 'You see, I soon had to start extending our goals because we simply did not have weather luck. This weather luck is a very important influence when it comes to wildlife recovering.' The wheat stubble becomes tangled tussocky margin and Francis waves his hand across the rough ground, as though casting seeds, and tells me it is 'high-quality habitat tailored to the grey partridge. It took two years, when we started, to implement this across 7 per cent of Rotherfield, and the trouble is, if you would measure everything out, across Britain you find just 2 per cent of farmland is this high-quality habitat.' Just ahead of our feet, a silvery shrew breaks, its tail darting along behind it as it dillics this way and that, over the parched earth, before disappearing down into a crack in the mud. Francis stoops, picks a pale purple flower, and holds it out for me. 'Here we have our good friend phacelia, who we all seem to like so much.' The air peels with the keening cry of buzzards and, as we walk, I try to count them, while tearing at the purple plant with my nail. Francis has fallen behind and is crouching among the stubble. 'Look, we have buckwheat and linseed.' I turn, walk back, and stand above him. 'You see this?' He holds his hand out towards a group of petals that are growing taller than the flowers around them. 'This is lupin.'

Out to the west a small plane appears over the top of a large beech wood, and Francis, still on his haunches, turns in its direction. He stares at it, following the droning hum in wide-eyed silence as it passes low overhead. I ask him where it's going, feeling it would be rude to let the moment go unacknowledged. 'I haven't a clue,' he replies, disinterestedly, before turning back to the lupins. 'What this is, is enhanced overwintered stubble. What you must do after harvest is leave it until the following August, so it is in essence taken out of the farming cycle for one year.' He speaks slowly, looking up at my face, seemingly trying to decipher whether I understand the significance of what he's saying, before he stands and walks on down the hill towards where the bones of a stream run between the fields. 'The point,' he continues, 'is to provide suitable habitat during the winter for the partridge and some food to forage into the summer. Here, across Rotherfield, we have 20 hectares.' As we go, Francis tells me that every plant he has chosen to add to the winter stubble has been selected for a specific purpose. 'When you look down,' he gestures to the ground, 'there is still plenty of habitat, but by February and March, the partridges, they need more. This, it is greener so there is something to eat and it's taller than the average stubble would be, and then in spring it attracts insects so that the partridges can feed their chicks.'

As we wander back beneath the telephone wires, five greenfinches, as one, flutter up into the sky, wheeling above the earth and beneath the raptors, growing smaller and smaller, until they disappear completely into the glare of the autumn sun. Francis runs his finger up and down his outstretched palm, telling me that the strips all around us replicate the margins and divisions that existed before agricultural intensification. 'In English I don't have the word, but you could say fallow. Where in the past people would cultivate one field and allow one to rest rather than to have all in production.' In front of a small clump of fruiting hawthorn, we turn back to face the way we came. 'Of course in the past,' he continues, 'stubbles were left over the winter and the crops were planted in spring, but we have less of this winter cover now.' We stand, looking back at the land lying still and I suddenly feel aware that across the country the earth is being torn up. Not long ago, winter crops were a rarity.

While they produce higher yields, they are also at greater risk from diseases and insects, meaning they require more spend on herbicides and pesticides. It was only when these became widely available in the 1960s that technology sated our hunger and the seeds of a new agricultural chapter were sown. Francis clenches his fist. 'These days, as soon as it's harvested, the field is sprayed off with chemical, ploughed and a new crop is drilled.' Lips puckered, he raises his arm, and makes a noise like a bird scarer going off while punching his hand up into the air. 'I see it all happen within one week.'

Francis steps forward, reaches into the thorns, picks a red berry and rolls it around on his hand. Somewhere in the branches, a bird is singing. 'You see,' he says, shaking his head, 'sometimes you don't even need a whole hedge to save farmland wildlife. If we would only leave just a clump of shrubs. This is last-resort cover during the winter and the pairs like it in spring as well.' He walks round to the other side and a pink-winged flash passes my feet, the fleeing falsetto. I hear myself speaking as a small boy – 'but this can't be enough to hide from a sparrowhawk?' Squatting on the other side, Francis makes a gap among the thorns and his face appears, peering up. 'Yes,' he says, excitedly, 'a bush like this can save the life of every single bird in a covey all winter long. Actually, I took pictures when I was young.' He springs back round, kneels down, and reaches into a gap. 'So there was this buzzard, attacking the grey partridge covey, fifteen maybe, and they took shelter in a clump of bushes just like this.' His hands are clasped together playing the part of the timid birds. 'Under the dog rose. There is some dog rose here.' He moves back so I can get a better view. 'You see, the dog rose has these hanging branches.' A severe look has come over him. 'Dog rose as a cathedral, you know?' I nod, but it must be clear I don't, so he says it again. 'Dog rose as a cathedral, hanging'.

Francis leaps to his feet and his hands go from being partridges to the wings of a raptor from long ago. 'And the buzzard comes.' He rushes round to the other side – 'and the buzzard leaps on top of it but he cannot get in.' Watching Francis, my thoughts are drawn to those great tribal storytellers like the San people who spend hours acting out intricate performances to conjure up the spirits of the animals they share their world with. 'The buzzard then jumps down on this side and what did the partridges do – what does he do?' Francis leaps round while sucking air over his teeth. 'The partridge, ever more calmly, he walked underneath the shrubs. It was like that and in the end, the buzzard, he didn't get even one.' He stands, runs his hands through his damp hair, then turns to me and smiles. 'Sparrowhawk, okay,' he continues, 'yes, that would be a bit different of course but the partridge could go in here. You see, where I have planted this privet?' He clutches at the leaves to show me how dense the growth is. 'The sparrowhawks, they can't get . . .' but he stops and pushes his arm into the thorns before reaching his hand up to me. On his palm lies the speckled feather of a partridge, cold grey running to a rich chestnut tip. I run my fingers over the barbs and then fold it up in a receipt for safekeeping in my pocket.

The field rises and Francis walks ahead. 'We do sometimes see grey partridge at Rotherfield,' he assures me, calling back, 'even if today they are extremely hiding.' At the top of the slope, he gestures to a patch of sun-scorched grass beside a raised strip of flowers and then slumps on the ground. I sit across from him and he tells me that last year he became extremely angry when he took a group in their mid-twenties for a walk across Rotherfield to show them how they've created a habitat where farmland birds thrive. He holds his hands up, thumb and fingers pressed together, as though trying to cast a crocodile shadow on a sheet. 'Almost as soon as we set off we saw a big covey of grey partridge coming out of a clump, huge covey, truly amazing,' he flattens his fingers and moves them through the air like partridges catching the breeze. 'They didn't even lift their binoculars. Honest, I'm not even joking.' Francis shakes his head and tells me he gathered them round and asked them what their favourite farmland bird is. 'Of course, they said corn bunting or yellow wagtail, one of them said kite. I just said, "What's wrong with you guys? This grey partridge is the second-fastest declining bird after turtle dove in Britain. You don't even dare to look at it."'

Francis tells me he hadn't really thought about it much until that point, but he later realised that the grey partridge's status as a gamebird has made it symbolic. 'I think for the average birder, who maybe doesn't like shooting, anything to do with shooting they sort of object, and the grey partridge, even if it's a bird and it's a pretty good bird. There's no other bird like that in Britain from its behaviour, but still they don't love it.' For a moment we sit in silence. Out on the edge of the fields, almost as far as I can see, a woman is walking and by her feet, nose down, a spaniel quarters the earth. As it runs, a flock of doves rises ahead of it and then wheels round towards us, wings sounding a round of applause in the stillness.

The summer bank to our left buzzes with life and somewhere within something is stirring. 'You know,' Francis continues, 'if you really love birds, how can you not love my grey partridge? It doesn't go and do some adultery stuff. It's monogamous. Its behaviour is fascinating. They stay as a family group all winter. That is better than most people now, I think.' A shield bug appears on my hand and I place it on the head of a sunflower. 'I read the other day,' he continues, 'that over 80 per cent, maybe 90 per cent of insects have gone in Europe in the past 30 years, almost the same as grey partridge, which of course need the insects to eat.' Hoverflies drift down the strip and by our feet, bees bumble in the nettles and weeds, green leaves beneath tall grass tiger-striped with the light of the sun. The area has been planted as cover and nesting habitat but also to provide food for the chicks. 'One month ago,' Francis tells me, 'this whole area was covered in ladybirds. All of it. I think too poisonous for partridge to eat, but the chicks, who will hatch normally on about 14 June, love to eat the larvae.' On our way up towards the lane, the hare that had been moving among the sunflowers emerges with nervous eyes. As Francis watches it lollop away across the stubble, he tells me I must remember that when we do what is right for the grey partridge, we do what is right also for the other farmland creatures.

The Harrow doesn't take card and Francis and I can only cobble together 5 pounds between us so we ask for a half pint each and then we sit out on the lane by the badgers' setts where the air is rich and damp. The glasses have handles, charming when pint-sized but too small for our hands at half. The hope, for the Scotts, Francis tells me, is that one day they'll have enough grey partridges again to shoot fifty, but until then they've said there'll be no shooting at Rotherfield. He'd never be able to pull the trigger on one of his beloved birds himself, he explains, but over the years he's come to recognise that with a few exceptions the only people prepared to pour millions into trying to save the grey partridge are those who have a distant hope that they might, one day, be able to kill them. For all that, he isn't sure about shooting in this country. 'Gone a little bit the wrong way,' he tells me. 'In Switzerland, every dead creature, at the end of the hunt, is laid out and music is played to thank the forest for its life. Fox has a tune, hare has a tune. Horns playing, music that hasn't changed for 500 years.' The grey partridge shares a tune with other small game and Francis can't think of any piece of music that would work for it as an individual species. 'It's not nightingale,' he says, shaking his head. 'It's a beautiful bird, but it has just ugly, horrible call.'

Above us, in the cedars, the rooks are beginning to gather. I think back to those young people who preferred the corn bunting and the kites and ask Francis if he'd have fallen for the grey partridge if he'd grown up in England. He runs his fingers through his curls. 'I like this. This is interesting question for me. I was very politically radical as a student, very loud. I feel perhaps I would not, but it's hard to say.' I remember, as he's talking, that there was a fiver in my trousers the other day and I push my hand into my pocket, past the partridge feather, in the hope of another beer. It's gone, if it ever was there, and it could have been a year ago.

Fingers clasped beneath the wooden frame and rough thumbs on top, Gerald Gray holds the picture of his grandfather up to the light. I look from the moustached face of the man sitting next to me, to the face in the magazine cutting. The nose is longer but there is the same kind curiosity in the dark eyes. 'James Henry Gray,' Gerald says proudly, 'you can see where that was. That was in *The Gamekeeper Magazine*.' Down at the bottom of the stained paper, a date is written in bold: 'September 1931'. I read it aloud. 'That's it,' Gerald replies with a smile, as though he's asked me a question and I've hit on the right answer, 'that's when it was.' He passes me the frame, then turns and looks away out of the small shed-window across the soft December fields.

The feature, from the 361st issue of a now-defunct publication, is a profile of one of the great keepers of the interwar years. The writer notes that Newsells Park, 'an estate any motorist travelling from London to Cambridge will have driven through', is fine grey partridge country and James 'being the man that he is, puts his whole energy into the task of partridge preservation'. One apparent frustration, though, was the emergence of 'the motor poacher', men with 'fine cars' and 'well-to-do' appearances who, according to James, were in no way like 'the old-time honest poachers' but were simply 'dyed-in-the-wool crooks deserving of no sympathy'. When old Gray wasn't too busy keepering or too exhausted from sitting out all night, watching for armed men prowling the coverts, he was a valuable part of the estate cricket team. The writer notes that Gray has four children, two girls and two boys and concludes that there is little doubt at least 'one of the boys will someday be portrayed in a prominent position in this journal'.

The writer was right. At fourteen, James's son, Reginald, started working under his dad, and 20 years later, Reg's own boy, Gerald, followed in his father, grandfather, great grandfather, and great great grandfather's footsteps, to become a fifth-generation gamekeeper. On the floor around our feet are pages and pages, torn out down the years, that record the lives and passions of the Grays. Gerald tells me his first memory is of being in his parents' cottage, one autumn day 63 years ago, and standing up on the kitchen table to watch a covey of partridges wandering busily through his mother's garden. The following year, his grandfather took him out in the spring and taught him 'how to discover eggs'. With a grey partridge, Gerald explains, holding one hand flat over the other, 'it covers its eggs, so you've got to look for a disturbed situation. Something which is not quite as it should be.' There is a sense of magic in his voice. 'If you look to the bottom of the hedgerow, in the grass, you're looking for a trail leading in. What you want to see is where a grey partridge might have snuck in for the last two or three days.' When Gerald was a boy he remembers his grandfather saying, 'There's a nest in there, Gerald', and he'd pop his head into the hedge and say, 'No there's not', at which point his grandfather would tell him to dig a little deeper. 'I'd scrape away the grass with my hands and scrape away and all of a sudden there was eggs.' After he says it, he laughs and laughs. 'It's just the way it is with grey partridge. It's just the way it is.'

For all that his father and grandfather were celebrated, not so long ago, in the pages of *The Gamekeeper*, Gerald believes they'd be lost in the modern shooting field where gamebirds in their millions are reared in sheds. 'Grandfather, in particular, wouldn't want to be part of it.' Gerald has a memory of standing, the three Grays together, looking over a fence to the neighbouring estate where they'd just started releasing. 'It was the school holidays and we'd just been collecting eggs. Grandfather just said it was disgusting. It should never have been happening, he said. If you can't get them from the wild then what are you doing?' He trails off and bites at his bottom lip before starting again slowly. 'That was their attitude. To rear birds and to put birds down was something that was, you know, it was something that was just not right.' Thirty yards away, next to the white cottage, in a kennel beyond the grass, three black Labradors start whining. Gerald, without looking up, hisses between his teeth and the dogs fall silent.

We are 5 miles south of Swaffham and 16 west of Norwich, in the middle of a 4,000-acre estate where Gerald was headkeeper for three decades before retiring just over two years ago. There are no signs to Hilborough, but you know when you've got there. The landscape contracts and endless featureless fields become a patchwork of stubbles and wildflowers, glimpsed through gates between sprawling thorny hedges bound up in dog rose and bramble. When Gerald was a little boy, most of Norfolk would have looked like Hilborough and most of the shoots would have been similar to the one he ran, in the same style that his grandfather and father ran theirs, but beyond the boundary things are different now, and like the grey partridge, the wild bird keeper is a dying breed.

Gerald glances at my hand and looking down, I see that a lady-bird, confused by the 12 degree mid-winter morning, is crawling sleepily along a veiny ridge. In the 30 years that he lived in the headkeeper's cottage, 300 yards along the lane, the estate only ever managed to shoot one year in five. Creating the right habitat for the birds and controlling predators sufficiently was always complex, but the decision on whether or not to shoot was a simple one. 'At Hilborough,' Gerald tells me, 'you must have at least 250 pairs of partridge, counted on the ground in the spring, because I know they will produce over 1,000 young. Now, if we got 1,200 young, I knew we could shoot 200, which would leave you at least 250 pairs again to produce chicks the next year.'

On a table to Gerald's right are two photo albums and he opens them across his lap. There is picture after picture of men standing next to each other, tailored tweed breeks, cigars and shotguns in fine leather slips. They seem of little interest to him and he flicks on through before he stops and passes me the album. Set against a backdrop of cloudless blue, above a beech hedge turning for autumn, 13 grey partridges, wings set, are spread out across the sky. 'Sometimes, the gentlemen, when they come, tell me they used to have grey partridge on their shoots,' Gerald says with a smile, 'and I always turn round and say to them, well, did you used to have red legs and they say no, that was before we started putting them down and I say, well that's why you've got no grey partridges because you shot them out.' He laughs despondently and shakes his head. 'That's the truth of it. The real reason there aren't any grey partridges in certain areas is because over a period of time, it doesn't take too many years, you just take that population away by shooting them.' The trouble, he continues, is that when grey partridges have a bad year because of

217

predation or poor weather, men like his father and grandfather would tell their boss they wouldn't be shooting. With reared birds, though, you can put down as many as you want and being unable to differentiate between the vulnerable greys and the shed-reared red legs, Guns stand in the field, merrily knocking down every bird that flies by until eventually the greys are gone, the red legs are all that remain, and it's only those like Gerald who realise that an almost irreparable act of destruction has been done. He takes the album back and continues leafing through, pausing for a few seconds, every five or six pages as though a face, a dog, or the way the sun is shining lights up a memory. Towards the end of the smaller album he stops at a picture of a man in his mid-thirties, mouth twisted and with tired anxious eyes. 'That's Nick. He was the headkeeper at Holkham, but when they had a little bit of a problem with birds of prey and things like that, he had to move away.'

Before rearing became ubiquitous, Gerald can remember butchers, right up to the 1950s and '60s, paying two-pounds-a-piece for a grey partridge, whereas now there is such an abundance of released red legs, that some game dealers charge shoots to take birds away. With the value of game having declined, the appeal of poaching has waned, but there was a time when it was an almost constant concern. Gerald scans another article, running his finger along the words. 'I'll just find it here. Yes, so it was my great grandfather's brother was killed by a poacher, shot. Actually, we lost two of the family to poaching in the 1800s. Another relation of my great grandfather, he was beaten almost to death up near Heydon. Then they pushed him under a train.' Outside on the dappled lawn, the sparrows are singing short trilling songs. 'There was the honest poaching then, that was just to feed the family, but you see there was always the possibility that if the person who did it got caught, they'd have been put in a boat and sent to Australia.'

The dogs are up and barking and Gerald pops his head out of the door to look to the gate. 'It's Mr William,' he says brightly then sets off over the grass. On the gravel leading up to the cottage, blonde hair and camouflage jackets, two little boys run up the path in front of their parents. 'What do we say to Gerald, Hugo?' their mother calls after them. 'I say Merry Christmas, Gerald!' the bigger boy shouts in delight, bounding towards the old keeper with a parcel wrapped in bright red packaging, his small brother toddling at his heels as fast as his little boots can go. 'The same to you,' Gerald replies, as the boy reaches up with the present.

Tweed cap, deep-set dark eyes, big cheeks and a big mouth, William Van Cutsem stands tall in the garden. 'I'd have liked to have seen some drives fly a bit better yesterday,' he says thoughtfully, 'but it was good, in general.' For the next few minutes, Gerald nods as William takes him through a plan for the current headkeeper to plant more hedges in the hope of showing better birds, until he is interrupted by his wife shouting across the garden. 'Boys, what have we said? We don't throw stones at Gerald's fish.' In her French wellingtons, the mother strides across the grass and scoops up her sons who have been lying on the edge of the pond, throwing pebbles at the creatures swimming below them. 'We ought to, I think,' she says to her husband in exasperation, but he is still talking partridges. The boys run towards the gate and out onto the lane where the smaller one trips and begins to wail. William smiles at Gerald then turns and follows after them.

It is the eve of the winter solstice and by the time we are down the lane, beyond the oaks, standing in a gap in the hedge looking out over sandy breckland, the sun is starting its short descent. In the middle of the field, beneath a kestrel hunting shrews, there is a long strip of rough grass and wildflowers. Gerald tells me that many decades ago, long before his time, they dug pits in the middle of the fields to get down to a layer of clay which they would extract and spread over the soil with a horse and cart to stop windblow. 'Maybe there's a rabbit or two in there,' he says hopefully. 'Hares can live in that ground where the pit was. A roe deer might live in it, but the nice part is that if there's a pair of grey partridge in there, they have somewhere they can stay all year away from the tractor and produce chicks.' A soft wind has picked up and I ask Gerald, as I watch the kestrel sailing back and forth like a kite on a string, if it's possible to run a wild bird shoot while abiding by the law.

He nods deeply. 'It's got to be. If you look at it now: the few remaining grey partridge shoots that are doing well. This is going well. Arundel is doing well. Up in Northumberland at Alnwick Castle. There's an abundance of predators. We all work very hard to control foxes, which we can do legally of course, but there's no way this stuff flying in and out is.' He pauses as though wondering how far to go with it. 'It's an impossibility.' We carry on down the lane and Gerald tells me it has long been a frustration of his that people don't understand that raptors like goshawks do much less damage 'as they might be portrayed as doing. They're lazy old birds'.

On the verge to our left, a young rabbit lies dead, its front legs twisted up in the air and its white belly squashed flat. 'Poor thing,' Gerald says sadly, 'just chuck that in.' I stop and grab at its neck between my forefinger and thumb. The grey fur is sodden with dew and its body hangs stiff. It tumbles out from my reluctant hand as I throw it and lands with its back legs in the hawthorn and its head on the grass, sticky eyes fading to grey. 'That'll do,' Gerald continues, 'the rats there'll get that.' Looking down, next to the rabbit carcass, I see a narrow run disappearing into the hedge, earth worn away by hundreds of night-time claws. We cut left through the fields and

Gerald explains that he thinks 'rats is a much bigger problem than people understand. You go out at night, you'll see them in places you never expected. In the barns, you see them in holes you didn't know you had. You'll see them climbing in the hedges. You'll see them in the trees. They'll take any nest they can find in spring. They're capable.' He purses his lips and whistles in amazement. 'I've even seen them rolling eggs along the fields, and if they disturb a partridge with chicks under her at night, it'll grab 'em.'

Somewhere up ahead in a strip of mustard that runs back to the cottage, greys are calling and the noise drifts back to us on the breeze. 'It's just this time of year,' Gerald says, as though telling me a secret, 'in the last few days before Christmas, when they might start pairing up.' He gestures to another strip. 'Some over there that I can just hear might say to these ones, we know you're there, we'll come across. Particularly if there's a bit of weather on the way.' All around us, beyond the estate boundary, there are large commercial red leg shoots and I ask Gerald how many birds they would account for in a season if Hilborough went the same way. 'You could shoot 5,000 easy,' he replies. 'Rather than these strips you'd have big blocks of maize, 100 acres each and you'd wipe out your grey partridges in two years, definite. If you got to a third year you might find an odd one, but in that first year you'd knock them for six and in the second year you'd wipe them out.' He tucks his hands into his pockets and starts laughing. Whether he's laughing at the fragility of everything he's ever worked for or whether he's laughing at the blind destructive madness of it all, or something else entirely, I don't really know. 'It's the business side to it that's changed the way it's been done,' he says gravely. 'Shooting wasn't ever something to be made money out of. That's a totally different world.'

The sun is fast falling away towards the fields, turning the water in the tractor ruts gold, and the partridges are getting louder. As we come up to the end of the mustard, Gerald says, 'truth of it', he doesn't think anybody would have the passion for grey partridge that's needed to save them if you weren't able to shoot a few. 'There they go.' By the time I look up and follow Gerald's line of sight, the covey has curled away over a knapped flint wall and disappeared into the light. The dogs start up again and Gerald calls for them to be silent before telling me that he's due to go down to Swaffham on Christmas morning to pick up his old dad. It's been a difficult year for Reg. Back in spring, at 96, he got engaged to a 'lady friend who lived in the same retirement home' – Gerald smiles and shakes his head as he says the words. She was a year younger than him, but before they could get a date fixed for the wedding she died.

Elbows resting on the cold steering wheel and unshaven chin in his hands, Frank Snudden smiles and tells me patiently, for the third time, that it's the chestnut horseshoe I want to be looking for, 'chestnut horseshoe on the chalky grey breast'. He passes me his binoculars and I try again. The spring sun broke an hour ago at five but the frozen fog is only just lifting and I can't see clearly enough to make out the partridge in the chicory. 'I'd say she's about 10 metres to the side of him,' Frank continues. 'She's on the left sort of sitting down but what I can do is I'll flush 'em.'

When he turns the key in the truck's dashboard the petrol engine stutters out into silence and the sound of dawn rises. Across the field, in a row of maples, a song thrush is singing and a wood pigeon calls to its mate. Like an animal, Frank moves quickly and silently over the dewy earth. First foot in the cover and the birds break, the cock clucking in high alarm as they swing round in front of me, one on top of the other, no more than 10 inches apart.

He watches them until they disappear out of sight and then he walks back laughing like a boy. 'See that, see how the greys stick together? Pretty cool. Pair of red legs, one would go one way and one would go the other, but greys'll always find each other in flight.' He reaches behind his seat and pulls out a muddy white map on a clipboard. It takes him a moment to find the exact part of the farm we're on, then he marks the spot in blue. For the past three days, six-field plots at a time, Frank's been out at dawn counting and recounting every pair of greys on the farm. 'So far,' he tells me, raising his voice over the sound of the engine as we drive away, 'they've been showing as good as I've ever seen them, but there's still five mornings to go.'

Five years ago, when George Ponsonby decided to employ a wild bird keeper in order to establish a grey partridge shoot on his Gloucestershire farm, the total population across the 1,700-acre holding had fallen to 16 pairs. Frank was just 25 then, but he'd served his time as an underkeeper at Arundel, the Duke of Norfolk's Sussex estate where England's grey partridges are at their best. 'Thing is,' Frank tells me, raising his voice over the engine, 'dear old George isn't a rich chap on the level that most people involved in grey partridges are. He's just a farmer who's super passionate. It took three years to get the numbers back to a level that we could shoot just only a handful and he's paying for all that labour.' As a young man, George started off with a 200-acre family farm and every bit of land he's taken on since is rented. 'He's got to make money,' Frank continues. 'There are some grants for habitat but it can be frustrating and there are arguments. I'd like to have more habitat but farming here could never take a back seat.' We turn left out of a wet wood onto a field of stubble and Frank pulls up to mark down another pair of greys in a strip of kale. 'In an ideal world, I do believe we could have around 100 pairs, but we haven't got there just yet.' He moves his head from side to side as he says it, as though weighing up the ambition. 'We were right up to 72 last year, but so much depends on the weather.'

Two fields on, where stubble gives way to grass, Frank slows down and glances into a hedge. Seven inches above the dark earth, a braided metal loop hangs from a wire. He leans over to take a proper look then accelerates away. Beyond the next hedge, another pair of greys takes flight from beside a water trough, then settles a short way out in the middle of the meadow. 'That's the classic pose,' Frank says, switching off the engine. 'Do you see? The hen's always sat down like that and he's always up looking round and protecting her.' He puts the binoculars down between us and points to his neck. 'He's got a greyer neck, chalky grey, and a sort of chestnuty head and she's kind of like a light grey.' It's a picture Frank sees thousands of times a year but he talks about the two birds in the grass as though it's a scene he's been waiting to see for a very long time.

Out in front of us, the sun is casting Lechlade in pale watery light and the spire on St Lawrence's is shining above the last of the fog. I offer Frank a caramel wafer. He talks as he chews, eating as though he hasn't eaten in days. 'Shooting is a funny thing, you know. It is a funny thing. You work your big bollocks off and it takes over your life. I think about trying to keep these things alive all the time. Then when it goes right you just drive them over a hedge and a load of rich people blast them.' I pass Frank another wafer and he holds it in his mouth as he turns the key in the truck and pulls away. He says he fully gets it that people might not like the idea of shooting a red-listed bird, but he reckons that on land where greys are being prioritised some shooting is essential to restore the population. 'Thing is, obviously a cock grey will get more dominant the older he gets but his hen, because they mate for life, will have smaller and smaller broods the older she gets. What you have is a bigger territory and a smaller hatch so you don't want hens lasting four years.' Twice he asks me if it makes sense.

As we head towards Southrop, the village that lies in the middle of the farm, Frank stops the truck and tells me there's 'another snare needs checking.' I follow after him to the edge of the field and he kneels down in the dark earth. 'Honestly,' he says, unsnagging a thin loop of metal, 'if snaring was banned this would have to stop. A lot of lads think they're heroes because they drive round shooting foxes all night. That's fine, but they're not achieving anything.' He runs his fingertips around the wire then pulls it tight, then loosens it, then tightens it again. 'Foxes only really do damage when hens are sitting in May or June, but obviously all the vegetation and crops are up by then so you can't shoot them. So honestly, if snaring was banned, I'd leave wild bird keeping. You couldn't do it.' He stands up and kicks at a bit of fox scat with his boot. 'I caught one here last week, that's why it's all a bit knackered.' In my mind, I see the shitting fox with the noose around its neck and I ask, before thinking about it, if it was panicking. Frank shrugs at the question. 'They're pretty pissed off, inevitably. I would be, but there's two points of swivel so it's not all frayed around their neck or anything. He was just sat there.'

Back at the truck, I fetch my flask of tea and we sit on the bonnet. After a while, Frank tells me that it's not like you see in those pictures 'where they're hung up on a bit of barbed wire or they're hanging over a bridge. It's just the most efficient way. Without snaring, it would be 100 per cent over for the partridge in England. Trouble is, what it comes down to is snaring, and Larsen trapping for crows and magpies are the most . . .' He trails off, looks down at the ground and then bites on the knuckle of his thumb in thought. 'Effective,' I suggest. He nods, 'Yeah, effective but it also begins with a "c". It's what people don't like.' Under his breath he tries 'commercial' then closes his eyes. 'Controversial, is it?' – Frank laughs when I say it. 'Controversial, that's the one. Snaring and Larsen trapping are the most controversial.'

By the time we pull out onto the road, heading east, the sun has burned through the milky fog and the sky is bright. For a couple of minutes we drive in silence and then, in a tone I haven't heard him use before, Frank tells me that 'it is a strange one with all the killing'. He hesitates and then says it again, but differently, and I'm left wondering what he meant by it and if it meant anything at all. 'It is a strange one but like, at a time when everything is on its knees except predators, I don't know why people aren't more accepting of control. Crows, magpies, foxes, they are some of the biggest threats to our red-listed birds. I don't know why people is so dead against it.' Frank asks me if I think that in France people are better at just letting others get on with their lives rather than looking to fall out over everything. He seems disappointed when I tell him I've only really been to Paris twice and it never came up.

In the village, a man in a fur-trimmed parka is walking something like an Afghan hound but otherwise the main street is quiet. On the left, just before the bridge over the Leach, water still high with last week's rainfall, there's a large pub covered in Virginia creeper. 'There's better to be fair,' Franks says, as we pass it. 'That's a bit too . . . I don't know. It's full of yuppies. When I used to go in there and ask for a few pints of cider, we'd get looked at funny.'

George's fields beyond the village are wet and Frank knocks the truck into four-wheel drive. 'Even one of the tractormen didn't used to like the snares when I first came,' he tells me as we pass another, 'but that's classic. I think people see stuff, they don't like it but they can't relate to the bigger picture. He moans about my snares but he'd happily go out in his tractor and work the field in the middle of the night and squash all the hedgehogs.' Part of Frank's trouble with foxes is that George farms in the heart of hunt country. 'They're fanatical about it round here, so they want the foxes to hunt, which makes my life hard work and I fall out with a lot of people over it. Me and George have seriously fallen out with our neighbours.' Ahead of us, on a shallow rise, a partridge breaks from a strip of wildflowers. Stuttering a scratched-up song, it casts away over the field, then flutters down, chestnut head jutting in the wet grass. 'They shouldn't be doing it,' Frank continues. 'What they're telling me is that I can't do something legal to help a red-list species so they can jump around the countryside, making a mess, breaking the law. That's what it comes down to.'

He stops the truck at the top of the slope. 'Did you see that? The flash of silver, they call it, because they've got white under their wings, and obviously it looks silver in flight so it's the flash of silver.' He shrugs and laughs, seemingly suddenly self-conscious about his excitement. 'That's a single cock there. His hen went off ahead of him. He'll be pissed off about that. If you listen carefully, you can hear her too. Can you hear her?' I listen but all I can hear is the male cheeping in lonely agitation. 'When they're together, it'll only be the cock that calls,' Frank continues, 'but when they split up they call to each other.' We sit for five minutes and he tells me I have to be patient. 'That's half of it. If you're patient, we'll see them meet up.' The little bird, chest puffed out, strops in the grass. We eat the last of the wafers and Frank asks if I always wanted to write. 'I'm not sure,' I reply. 'I did an interview actually to do gamekeepering at college. I got a place, but I also had a place to do literature at university. I sort of got talked out of keeping.' Frank laughs and says he always thought they were sort of the opposite. 'My missus does all my paperwork, all my admin – my wife now, I should say. She's bloody good, she's really handy, I'm dyslexic so she writes all my stuff.' Out in the field, the cock's calling rises and then he takes flight, beating his wings for 30 yards, before dropping into the cover where the two birds settle in silence. 'In truth,' Frank tells me, 'it doesn't matter how many pairs I count this spring. If it rains throughout June, relentless, if it's cold and wet and if the chicks haven't hatched, the hens will just sit on the nest. They won't leave them and then they all die. They all just die sitting on those eggs. It hasn't happened yet here, but it will.'

Boyish at 59, Paul Portz, Michigan-bred and London-made, looks out across the inky sky, blinking in the rain. On the wind, a flag cracks, and through a tall blackthorn hedge, rising up in front of us, the outlines of 42 men appear, shoulder to shoulder marching through the mustard. 'Portz' twitches as a charm of goldfinches twists above our heads and then the flankers' whistles start going and grey partridges begin to break. Cutting behind a large oak, leaves mostly gone, a bird curls round on the cold easterly. When it's 40 yards out, Portz pulls his gun into his shoulder and tracks it for a brief moment before pulling the trigger. Dog at his feet, 10 yards from where the bird falls among the barley stubble, George Ponsonby stands smiling.

Four long Ukrainians

Farewell to the bushy clump close to the river
And the flags where the butter-bump hides in for ever;
Farewell to the weedy nook, hemmed in by waters;
Farewell to the miller's brook and his three bonny
  daughters;
Farewell to them all while in prison I lie –
In the prison a thrall sees nought but the sky.

John Clare, 'Farewell', written around 1860,
a poem on incarceration and lost waterlands

In the soft mid-autumn air, legs dangling beneath its nutty grey body, a crane fly flutters round and round before drifting down onto the roof where it disappears against the water reed. 'Fetch up four long Ukrainians, Bobby, please.' The words come out of Chris Dodson's mouth as rich as the Fenland soil running away into the damp distance. Footsteps sound on the ladder and a gawky boy appears, eyes downcast and two bundles apiece on bony shoulders.

'Reed's been coming in since about the 1970s,' Chris tells me, as he slices through the twine with his billhook. 'Certainly, I'd put it that 90 per cent of the water reed being used today by thatchers in

England is imported. Mad really, thinking that Ukrainian's cheaper than Norfolk when Norfolk's just an hour up the road.' Chris talks and works at the same pace, his rough fingers running slowly across the bundles, letting them find the shape of the roof as he tells me about how different it all used to be back in the school holidays when he was a boy. They'd set off, him, his dad and his grandad, trundling over the Broads in the lorry from Cambridgeshire to the north Norfolk coast. 'It always started at four in the morning in those days. It was absolutely marvellous getting up that early and we'd always have something to eat with us in the lorry. Grandad said that was the most important thing.'

Chris can't quite remember exactly when it was but he knows it was on one of those cherished trips, when the sun was breaking over the Broads, that he first saw the hunched silhouette of a bittern stalking fish at the water's edge. Even then, the marsh that three generations of Dodson wound their way through, along the banks of the River Ant, was a relic of a lost landscape. Hundreds of years ago there were thousands of square miles of marsh and bog across England, but today just 10 per cent of that wetland remains and farms have been carved out where the male bittern's velveteen boom once sounded in spring.

Efforts to drain East Anglia began all the way back in 1630, when a group of 'Gentlemen Adventurers' led by the Earl of Bedford cut two rivers through the Cambridgeshire Fen, hoping that the water would run off into the Ouse. The scheme was met with ferocious opposition from eel fishermen, wildfowlers and reed cutters whose families and communities had forever depended on the marshes for their livelihoods. In time, those who fought back came to be known as the Fen Tigers and for centuries they sabotaged drainage efforts and rooted out their enemies by setting fire to the reeds. But ultimately, in the eighteenth and nineteenth centuries, the mass deployment of steam-powered pumps became too much and the only way of life the Tigers had ever known rushed away as the water was drained from the land.

During periods of great hunger, bitterns were a source of fatty meat and the impact of the loss of vast swathes of habitat was compounded through hunting. By 1886 they had disappeared from England, but in the early 1900s word started to spread that their voices were being heard once more on the Norfolk coast. The first time the species is documented to have bred again in Britain was in 1911 when Emma Turner, a naturalist who spent her summers living on the Broads in a houseboat, took a photograph of a young bittern near Hickling. As is usual when they know they're being observed, the chick's bill is pointed straight to the sky in order to look like the reeds all around it. Turner titled the picture, 'Striking Upwards'.

In the first half of the twentieth century, much of the reed, across wetland that remained, was maintained to supply thatchers. The traditional rotational cutting cycle created patches of reed of varying lengths that were kept free of ever-encroaching scrub. The result was a perfect habitat for the secretive bittern, with fish and amphibians providing rich pickings among lush cover. 'Don't get me wrong,' Chris says, as he spreads out the final bundle, 'I like using Norfolk where I can. It provides jobs for boys like Henry Randell and we're looking after our bitterns, but it's the labour cost. They must be paying them Ukrainian chappies two cans of beans a day.' He reaches for an old varnished paddle with copper rings nailed into the wood and a long warped handle. 'It's a shame really, I've got to say.' The Randells, Chris later tells me, are a crab fishing family who have supplied the Dodsons for four generations. Throughout the summer, young Henry catches crabs, which are boiled up and sold by his grandad out of a shed on the coast road to Blakeney. Then when winter comes and the crabs scuttle out into the deep, too far for a small boat, he pulls up his creels and, like his father, grandfather and great grandfather before him, Henry turns to cutting reed.

'The basic tools haven't really changed a bit', Chris says merrily. 'This is a leggett. All different varieties but most of them look like this. What we do with it is we dress the butt end of the reed. We're knocking it up into shape.' As he works away, dust and seed fly up then float down around him in the still air. 'I can use it for whatever angle I want, different varieties, you've some with crossway lines, some with horseshoe nails in, just depends on the thatcher. I can turn it round this way. But if I want I can also turn it and use it horizontally as well.' He puts the leggett down, then stands and walks along a plank to pick up his hammer. I ask if he made the leggett himself and he shakes his head and laughs. 'There is a chap actually who still makes them, but Grandad made me so many that I won't ever need to buy another.'

Over the road in a hedge, hawthorn dripping with dew, a blackbird is singing. We stop and listen, but it falls silent then flits away into an oak, late autumn leaves burning from red to brown. 'This was a bakery once,' Chris says, casting his gaze across the roof. 'At that point, before the war, there was thatch all across East Anglia. It tended to be a one-up one-down kind of a job in those days. It was a poor man's roof like a farm labourer's cottage. Sometimes, if they couldn't afford to rethatch they'd just put the wriggly tin on there and throw the thatch over the top, nice and cheap.' His hands are full of splinters and he sucks at them between thoughts. 'You still see them like that sometimes and bugger they're black underneath, because that's been on since the war. Where are we, that's 70 years ago now. Sorry, 80 years ago, it's all got a bit mucky, I've got to say.'

Chris is in the middle of telling me that most of those cottages were torn down in the 1950s, the incentive for maintaining bittern habitat disappearing with them, but he trails off mid-sentence when he hears the clop of unshod horses' hooves. 'I'll wait till they go past,' he says, in a wide-eyed whisper, no quieter than his normal voice. 'One thing you can notice, if you're higher than a horse, they go banzai.' Two coloured cobs, side by side, plod along the verge, twisting their ears. As they draw level with us, the horse on the outside starts to pull its back end round and the lady sitting astride brings her crop down across its withers. Chris inhales sharply as the animal lurches forwards.

When the horses pass the far beeches, Chris reaches for his hammer and holds it out to me. 'A decent hammer is a must, you see. You can't have a toppy hammer that's going to just tickle it. You need something with a bit of guts.' He kneels on the roof and knocks a metal spike through to the rafters, head cocked like a collie dog listening, while he tells me there can't be more than a thousand thatchers across Britain now. To my ears, each thud sounds the same but it is only third time round that he says he's got it right. 'It needs to be tight enough that it's not going to fall out but it can't be too tight. Tight looks good, granted, but what happens if it's too tight is that rain will sit on the reed and it won't run down. You then have a capillary effect when the sun comes out and the water gets sucked up the reed. Terrible potential for fungus.' I ask how tight too tight is but Chris shakes his head. 'It's feel and experience. It's not really a spirit level sort of thing. It's not about measuring tapes. Every roof is different and it's just eye and knowledge.' As he pulls away a steel rod that's been holding the reed in place, he tells me that over the years he's only known about one in ten apprentices with the natural eye to go on and become a good thatcher. I look across the roof at Bobby, whistling to himself while working away with his leggett.

Chris passes a long sharp pin up to me. 'These spikes, or spits

235

you might call them, or crooks or hooks depending where you are in the country, these would have been hazel once, but lovely though it is, you've got the susceptibility to woodworm. You had roofs that would have been dandy, plenty of thickness still, coming off because the spikes got worm.' On the lane below us a tap-tapping sounds on the old bakery wall and I walk along the plank to look over the scaffolding rail. Shuffling his feet, an old man with a white stick held out in front him scrapes down the lane. Past the bakery, he taps along a row of red-brick cottages, and then onto the white-painted stone of the pub on the corner, lights off, curtains drawn and a polyester Union Jack hanging down over the door.

'Truth be told,' Chris continues, hammering in another spike, 'even the spars, made from willow, that twist along the top to secure the reed come from Eastern Europe now. Grandad had a little patch where he planted some, but he died two years ago and it had to get sold. Last summer, I nipped in and took a few cuttings. They've taken off really well.' Chris stops and looks up. 'Bobby, could I have six bundles of Norfolk, please.' The boy slopes past and climbs down the ladder.

Some 100 yards up on the other side of the lane, there is another thatched cottage with two cockerels fashioned out of straw on the roof. I ask Chris if it is one of his, but he shakes his head. 'When Great Grandad died, Grandad had been thatching the longest so he set the craft standard but then his youngest brother got involved and he cocked it all up so my Grandad, he went on with Jack. Then Arthur and Malcolm, they stayed together. That's Malcolm's boys did that. My cousins. They've got a different approach.'

Chris stands and looks across at the roof. 'It's fine. They like decorative ridges. That's fine.' It's not clear whether he's criticising or defending them. A rook flies low through the stillness then flares just before it reaches us, startled by two men being where men usually aren't. As Chris watches it wheeling high above, rasping in alarm, I ask him if he's ever seen a straw bittern on a roof. He brightens and takes his cap off before running his fingers through his thick reddish hair. 'There's two grains of thought behind finials, which is what you call them. One was that it kept witches away. The other, which I quite like, is that they say they used to put two hares on the roof and then when you got your final payment, you took one off so everyone knew who was good for the money.' It isn't something, though, that he's done for a while, ever since he and his grandad were asked to knock up a boar for a woman in Northamptonshire. 'This lady, she made such a fuss and we kept saying no and no. In the end it was my old dad who relented. He said here's some wire netting, boy, and some straw, have a go over the weekend. Well, deary me, Stevie Wonder could have done a better job after half a bottle of gin.' Chris shakes his head. 'It looked terrible, but Grandad, he helped me out and we put it up on a thatched gable window because we didn't want it on the main roof and the woman, she was furious. She just kept on saying it's got no bollocks, over and over. I had to go up and take it down and she just stood there in the garden below me, hands in the air, shouting it's got no bollocks. She was really quite a grand lady, I suppose.'

Bobby brings the six bundles of Norfolk in two trips and passes them to Chris who has climbed onto the ridge. 'Come up,' he says, 'and I'll show you the difference.' I clamber up the ladder and he hands me two reeds. 'See that. See how much thicker the Norfolk is on top. It costs more but I want to fill the back of the roof out and it'll do that quite nicely.' I snap the silvery silken husk off the top of the Norfolk and fold it up in a receipt from my pocket. From where we're perched we can see a long way west across the fields back to the road. On the same autumn day, hundreds of years ago, reed would have run for miles across the land and the bitterns would be hunting hard before the weather turned cold. Now the land is fields, and furrows scored straight as a die are ready to be sown again. In the distance, the cars are backed up, bumper to bumper, some accident unseen. One dead, I later learn, and thousands annoyed.

'Do they burn easily?' I ask, running my palms down the thatch around me. 'Best way to think of it is like a closed book,' Chris replies. 'You try setting fire to a book, very difficult when it's compact. There was a lady a few years ago, not far from here, who went out in the middle of the night and tried to set fire to seven or eight houses. She had a butane gas torch.' He reaches for his billhook and starts spreading the Norfolk reed along the ridge. I ask why, and he looks up as though he's never really given it much thought. 'Well,' he replies, 'I suspect she must have had a bit going on.'

Under the awning a white cat with blue eyes sits out of the rain on top of the pots. 'That's Thaw. Mrs Thaw.' Henry Randell smiles then jeers at the creature whereupon she leaps down and runs into the red-brick semi through the open kitchen door. 'Well normally, if you get a good year,' he continues, talking with his chin tucked into his green jacket, 'you can just put them straight back at sea, but then if you get a load of weather like last year they end up looking like this, shredded to bits. To mend this is probably a good half-day's work. It's never-ending.' Henry's red hands twist a length of blue rope round the battered carcass of a pot in front of him, an upturned fish box for a work surface. His eyes are fixed on the job and he doesn't look at me. 'That's another reason why you've not got many new people coming into it because they cost silly money to buy, about a hundred quid each.' He glances up at the jumbled stack. 'That's about three hundred pots there. You're looking at thirty grand just to go to work.'

Henry started crabbing the year he started on the reeds. 'I've done it since I left school when I was 16. I'm now 26 so I suppose that's 10 years.' He gives the blue rope a final tug and then starts laughing as though all we can do is laugh at the passing of time. The pot is thrown on the pile, he picks up another, and then staples away at the frayed meshing with a hog ring gun. 'My grandad's done it. He's done it for the last 40 years. He had reed beds at Thornham, at Salthouse, Blakeney, some at Cley.' He talks quickly, rattling through, but carefully sounds out the names of the marshes. 'We've gone smaller scale now with everything that's coming from abroad and what with getting pushed out, so we only now cut at Cley.'

For the past year, Henry has been having trouble with the Environment Agency, but he isn't totally sure why. 'That's the same as anything. Local traditions now seem to be less favourable. People want a pretty walk along the bank and some sort of nice scenery. They don't want to see my equipment down there or me I suppose, but I always stacked it away neat. It's crazy really.' He shakes his head and looks up. There is a white-haired lady walking past a burst drain on the other side of the road. 'Oh hello, Jackie!' She stops and looks across at us, squinting. 'Oh hello, Henry!' she shouts back. 'She lives up the road, that's Jackie.' Henry explains.

'If you called them up personally and asked them, they'd tell you they don't want to push us out, but they're stopping us carting the reeds and cleaning them. It's no different. It's crazy, but they'd never admit they're trying to stop us.' The white cat slips out through the back of the pots and Henry smiles then snarls at her like a tiger. 'I shouldn't even be saying this to you, but they haven't got a clue on the job. I mean we're limited like this with the wet weather. If I cut when it's wet, they rot. The reed is tied up so tightly in bails, bunched, 75 bunches to a bail. If they don't get no air round them, they just rot. Then Chris won't be very happy.'

The rain starts falling harder and big drops run down my ears and across the back of my neck. I squeeze in between the pots to shelter under the corrugated plastic. The day is warm and a faint fish smell hangs in the damp. 'The difficulty is that I need to get the reeds cut before the colt come through. About early April that new growth comes, so what with all that grief and now with this rain. We're in a bit of trouble, in truth.'

Every year at Cley, just before he stops cutting, for as long as he can remember, Henry has heard a bittern. By his reckoning he's pretty capable on his birds and, as he sees it, the cutting does them a world of good. 'We cut and then the bittern moves in, whereas where they don't cut, the growth and vegetation is so thick no bitterns can sort of really walk through.' I ask, having read it in a book, about the scrubby vegetation being called 'carr' in Norfolk but Henry shakes his head. 'That brown stuff, oh right, is that what they say we call it?' He shrugs his shoulders and laughs. 'I don't think it's got any use really, unfortunately.' For a few moments he works in silence, the sound of the rain drumming on the awning above our heads, but then he returns to the bitterns. 'That's lovely when I see that bittern, but people have to understand they're an extremely rare bird but you know what I mean, they're normally heading for the areas of reed that we've cut. It's crazy, really, what the Environment Agency is doing. What's in their mind, I don't know.'

I want to know what Henry would do if he wasn't able to cut reeds anymore. He laughs as though it's the stupidest thing he's ever heard and then tells me it's a bit like the crabbing now. 'Where I fish on this chalk bed at Cromer, they want to try and stop that. Perhaps if they all got their way, I would just go in an office like everyone else.'

Purple velour dressing gown, long grey hair, and a cat litter tray in his hand, Henry's neighbour appears on his steps and walks towards the bins. 'Morning,' he shouts over as he empties it out, 'or is it afternoon?' I look down at my watch. 'It's twenty past two.' The man curls his tongue round over his top lip then walks away over the grass, nodding in the rain.

In her final decade, Emma Turner started to go blind, but right up until the end she enjoyed visiting the wild places she'd loved, to hear the birds she could no longer see. In 1940, Turner died and her houseboat, the *Water Rail*, sat abandoned for 30 years until they smashed it up with a tractor bucket.

All around the bay, where the *Water Rail* was moored, the marsh is singing. Up in the willows two sedge warblers, not long back from Africa, are in full ever-evolving flow, and down in the reeds a coot whistles. On the pontoon, in the sun, the afternoon felt warm, but out on the water a hard spring wind is blowing from the east. Phil Heath, big grey moustache protruding down over his top lip and a blue hat hanging off the back of his head, sits with one hand on the throttle and his face resting on the other. He talks quietly, barely opening his mouth, and I have to lean across the deck to hear what he's saying. 'There was a hut on there then where she developed her photographic plates.' On the other side of the boat from me, with a Tintin haircut and little bright eyes, Nick Acheson perches. 'A fierce old trout,' he says, shaking his head and balancing his binoculars on his knees. 'A redoubtable woman, truly.' Nick talks of Turner as an embodiment of fierce brilliance and tells me that every conservationist does. His only regrets being that her short-eared owls were called 'Rabbi' and 'Shylock', and that note from 1911 when she wrote that the little bittern chick she found was very much like a 'little golliwog'.

By the time Phil started working on the Broads, in the early 1980s, it had been owned and managed by the Norfolk Wildlife Trust for almost 35 years. Prior to that, the marsh had belonged to Lord Willy Desborough, a wrestler, a Matterhorn-climbing man, a man who twice swam the Niagara rapids, the amateur punting champion of the Upper Thames, a Liberal politician, and a keen tiger hunter when he wasn't on the tennis court. Along with traditional tasks like organising Hickling's famous coot shoots, Desborough's keeper, Jim Vincent, was responsible for ensuring that the estate remained a sanctuary for birds.

As we drift down the leeside of 'Miss Turner's Island', Phil tells me that in his early days as a warden there were just six bitterns on the Broads and twelve in the country. Until the 1990s, when extensive research was undertaken, nobody really knew quite why the bittern was doing so badly or that what was needed was a return to the old ways. The findings, Phil explains, as he spins the boat and heads back towards Whiteslea Lodge, were relatively simple: for nesting habitat it was shown that bitterns require reed beds to be maintained, and for hunting it was found they need a mosaic of pools and channels rich with rudd and amphibians. Within three years of implementing a restoration plan, the number of booming males at Hickling began to rise, but Phil's feelings about their future aren't bright. In his working life he has seen fewer and fewer marshmen coming during the winter to harvest reed. 'It's just very hard for a chap with a wife and family to make enough money,' he tells me. 'So much so that the marshmen used to be charged to come onto the land but we now let them on for free.' If they stopped coming entirely, Hickling would have to start paying for the reed to be cut, but even that, on some parts of the Broad, may soon no longer be possible.

Phil eases the throttle towards him, the boat slows, and we turn up a dyke that runs back to the pontoon. Over the past few years,

Norfolk has been the wettest anybody can remember and while standing water is ideal in the short term for bitterns, it could eventually be the end of them. 'It's climate change,' he says, softening his voice further as he switches off the engine. 'Some of the water levels can be controlled on the marsh, but when it pours, just that rain non-stop, nobody can get on to cut.' Nick stops looking through his binoculars and turns to face me. 'So does that make sense, Patrick? What Phil is saying is that in an indirect way the lack of ability to harvest reed due to climate change could do for the bittern, because their habitat will just turn into scrub.' He smiles like a schoolboy sitting up at the front of the class, who has just got the answer right. Phil nods while uncurling a rope in his lap.

To the left of the boathouse, a large white lodge stands in the middle of a perfectly mown lawn. The windows are open and by the door an empty wine glass sits on a table next to a paperback with a broken spine. Somewhere inside there are four friezes, depicting the wildfowl and waders of Hickling Broad, painted by Roland Green in his vivid style. Green was the son of a taxidermist, and old marshmen who knew him recall that in the red-brick mill where he worked, he was always surrounded by stuffed birds. Nick springs up onto the pontoon and I turn to ask Phil if he thinks any of the successes he's had with bitterns could be replicated with other species. 'Wetland birds, maybe,' he replies, 'but in the wider landscape, no. I don't think there's any hope there whatsoever until farmers change and the human race becomes less stupid and arrogant and stops breeding, the crux of many things in this world.' I want to go into the lodge but it's leased by the Cadbury family and they're around somewhere so we slip quietly past, Nick striding on ahead of me, with his face turned towards swallows.

In 1952, Joyce Lambert, a local botanist and lifelong fan of Norwich City football club, changed the way Britain saw the Broads. Up until then, it was believed a natural process had created the lakes in northeast Norfolk, but using a small borer Lambert extracted core samples that showed the walls of the Broads are vertical. At some point, man had been digging. Following the revelation, Lambert dug into parish records where she discovered that between the twelfth and fourteenth centuries, in what was then one of the most populous parts of England, locals had sliced up and burned vast amounts of peat.

Nick is leaning over a wooden rail looking out across a reed bed that falls away to the water. He's travelled the world and has worked in conservation on every continent, but recently he's been rediscovering Norfolk by bike, setting out at seven each morning in search of the first birds he knew. Ahead of us, 100 yards out, a marsh harrier blows low then casts aloft on an uprush. Nick twists his neck to watch it fly above us and smiles as he imitates its cry. 'Forgive me for marching on,' he says, after the bird drops down over the woods to the west, 'but I'm on the lookout for bitterns.' Where the reed butts up against the farm next door, we turn right and cut along a narrow track. Beyond a dyke running red with fertiliser, a gaggle of geese wander in the six o'clock sun, their eyes following us as we walk.

On the right, running back towards the lodge, a new area of marsh is being created. Meticulously designed pools and channels have been carved out and reed grown elsewhere in polytunnels has been planted in the mud. There isn't enough cover for a bittern yet, but in time it's hoped they'll move in. I ask Nick what rewilders think of the careful approach at Hickling and he stops ahead of me. 'The thing is,' he replies, kneeling down to tie the lace of his boot, 'if you turn arable land back into healthier habitat, then fabulous.

It can only be better for wildlife, but if you stop managing a place like this for bitterns it'll simply all just become scrub.' We wander slowly on and he scrunches his face as though ordering his thoughts. 'Someone like Benedict Macdonald. He's not from a nature background. He's come to it quite late and he's picked up on this one grand idea to put all land back into wildscape.' Nick turns, one hand on his hip and another waving in the air, his voice becoming louder as he goes. 'Benedict Macdonald wants all nature reserves to be unmanaged. He wants you to just let out a bunch of horses and cattle. He wants pigs to wander round. Bitterns would be extinct within ten years in this country if you didn't cut reed.' He strides on ahead again and throws his arms up over his head. 'Ben is so zionist. Ben has discovered the answer.'

Thirty yards back down the path a man in shorts and wellingtons, big silver hair and a bright smile, is hurrying towards us, a clipboard held in both hands as though it's a bucket full of water and he's trying not to spill any. As he shouts 'Hello!' the geese lift and drift limply round over the marsh, more noise than flight. 'Hello young man,' Nick calls back. 'Honestly,' he continues, 'I get so arsey about it. Derek Gow and Roy Dennis and Ben Macdonald, and Isabella Tree to a more intelligent extent, they have this mantra that conservation has failed.' Nick raises his voice and smiles. 'That John individually has failed.' John nods vigorously. 'Oh, personally, yup, I've failed,' then adds, 'sorry, who've I failed now?' On our left, a bench in memory of someone who liked the marsh very much is set just back from the path. Nick sits on the arm and snorts, 'Rewilders, John.' The Hickling warden runs his fingers through his mass of hair, then shakes his head and bursts out laughing like an old man whose toddler grandchild has just appeared, little body covered in ink, and a pen in each hand.

John is only in his mid-fifties, but as we wander, he tells me it feels like he started working on the Broads so long ago, he can hardly remember how few bitterns there were in those days. 'There was one booming along at Martham, that was sort of the Broadland bird and then at Minsmere but once we got the water back up again, here on hundred-acre marsh, we had a boomer very quickly.' In many ways, bitterns are resilient birds. John has known the males to mate with up to eleven females in a season and the adults will happily see off marsh harriers, even eating their chicks when the opportunity presents itself. Like all the Broads, Hickling has a historic floodwall around it which meant the marsh was kept relatively dry, allowing cattle to graze. 'All we sort of had to do is let the water from the marsh back in.' John pauses and laughs at the simplicity of it. 'It's what we're doing here. Trouble is we've got these greylag geese eating the reed like mad.'

Set back among the thorns, a mill tower stands 40 foot high next to a cottage with boarded-up windows, and up ahead of us Nick is leaning on the gate. 'It was bloody wildfowlers in the 1960s who introduced the greylags,' he calls back. John shrugs in amusement and tells me that lots of people, in spite of them destroying bittern habitat, don't want any geese shot as they deem them to be a conservation success story.

When we catch up with Nick, I suggest in a half-formed sort of way that maybe rewilding has become quite political and Hickling feels a bit detached from that. Before I've come to any real conclusion he cuts in, 'But what's outrageous is that they're allowed to say all this stuff which is complete lies.' He jumps down off the gate and it swings behind him. 'They say we're enemies of nature, that nothing we're doing is creative or cutting edge. I am so tired. Look, let them come here and tell me this is not intervening for the benefit of nature, for the benefit of bitterns.' He breathes in deeply, then gestures across the marsh, reed lit up gold as the sun goes down. 'Are you as wound up about all of this as me, John?' The warden drums his fingers on his clipboard and pushes his glasses up his nose.

At the bottom of the red-brick mill, deadly nightshade grows among the daffodils. Until 1994, the Nudd family still lived in the cottage, a place where there'd been Nudds for 200 years, but they're gone now and it is all slowly decaying away. John runs his boot across a large piece of iron, half sunk in the earth. 'The casting sheared off on a calm night in about 1930 and the sails dropped off too. It was replaced with a steam engine that still drove the old scoop wheel to draw away the water.' The mill, one of many built across Norfolk in the late eighteenth century, drained the marsh and turned bittern habitat into grazing. The Nudds kept the dykes flowing, caught coots, looked after cattle, and John remembers old Gerald Nudd being the last man on Hickling to cut reed by hand with a scythe. They were part of modernisation once, employed to help improve a landscape where people had trapped eels and decoyed ducks for a thousand years. 'But in those days,' Nick tells me, 'there would have been wet grassland, little bits of reed bed, little bits of bog, little bits of woodland, a very wild landscape in which there was space for everything, a little bit here and a little bit there.'

Arms resting on the old wooden fence, we stand with our backs to the mill, looking east. If the winters didn't get any wetter and the men could come to cut reed, bitterns at Hickling would thrive, but John thinks it could all be over soon. When he's lying awake at night, in the old keeper's cottage on the edge of the Broad where the garden is full of coypu skeletons, he can hear waves breaking over the beach at Horsey Gap. It's 3 miles currently but every year it comes closer. 'The vast majority of bitterns are still only a handful of metres above sea level so that's the trouble. Climate change is the big worry.' In the shadow of a rotting wooden trailer, a young rabbit, no bigger than a fist, crouches stock still. For a moment we stand in silence, listening to the sky and then I ask how long they've got. 'If you went for a walk across the Broad in 50 years' time, would it still be much the same?' John laughs and the rabbit bolts towards the thorns. 'In 50 years' time, I'll be 106 but I would be surprised if the sea hadn't breached by then. It will go back to being a marine estuary, like it used to be in the Roman times.'

Beneath drifts of acrid smoke, a thin wind rises through the reeds. Somewhere back towards the village a fire is burning. Out in the distance, over Nudd Mill, the sky is dull and cold, but behind me bright clouds are billowing up across the setting sun, a white quilt sewn with a golden thread. On the marsh, a crow in a willow tree sits calling, and above it gulls fly back to their roosts, strings of teal head inland for the night, and swallows and swifts weave among them all. They said if I heard it I'd know, but when it happens it's only a faraway murmur and I can't tell. At seven the rain comes, a few light drops at first, rising to a steady patter on the tarpaulin tied down over the stacks of reed I'm sheltering beneath. Then, in front of me, no more than 20 yards out, three booms sound, a sweet hum, all electric and low. For 20 minutes I watch the reeds where I think I heard it, hoping it will walk out and stalk rudd in the shallows, but it doesn't and the marsh falls quiet and the crow flies.

When I call Benny, he says he's got two. 'One big. That's the better of 'em in truth and there's a smaller one there as well. Both Victorian.' I tell him I just really need to see them. I heard one at Hickling but actually I want to see one in the feather. 'Trouble is,' he replies, 'I'm busy all the while.' The following day I ring again and tell him I'm coming tomorrow. 'Trouble with that, I'll be out most of the day tomorrow and I'm busy, see, all the while.' I get halfway through explaining that I'll only be 10 minutes when I realise he's gone.

On the platform at Sheringham, a small blue-eyed boy stands in front of a bench, looking up at two men sitting side by side with a six-pack of cider between them. They smile down at him, burst blood vessels and black broken teeth, and he chats away. 'Going to Norwich,' he says, pointing up the Bittern Line, 'because I need new shoes before school.' The men look at his feet. 'What's wrong with those, then?' one of them asks, laughing. 'No,' the little boy replies, 'these are not school shoes. You'd get in trouble.' The boy's mother, who is on the phone, 30 yards down the platform, looks across and hurries over. She smiles at the men anxiously and they leer at her as she takes the child by the hand and pulls him away.

A warm autumn breeze blows through the elder bushes on the other side of the tracks and drops of light rain start to fall. I check the time. The 10.36 is fifteen minutes late. Across the road, in the supermarket car park, old people are fighting shopping trolleys. They push them one way and the wheels go the other. When the train eventually arrives, groups of excited children are up and waiting at the doors, all ready for a damp day by the sea. I step aside to let a little girl pass; she has a dinosaur on her jumper and a finger up her nose. Then I get on and find a seat beyond the stench of the toilets. As an announcement, unintelligible, comes over the tannoy, I glance back to see that the two men are onto their next cider now. Looking up, empty cans at their feet, they stare blankly towards me and the train pulls away.

At West Runton, the buddleia at the side of the platform is dying and three teenage boys get on. 'But I just don't get it,' one of them says to the others, 'how are they going to believe us if we say we were at the Shire Horse Sanctuary but actually we're in Norwich all day? You wouldn't spend all day at the Shire Horse Sanctuary.' The boy with prematurely thinning hair calls him a dickhead – 'you're a fucking dickhead, Harry, with your mask still on' – and they fall into agitated silence.

Trains first ran to Sheringham in the 1870s but at that point it wasn't called the 'Bittern Line'. Initially, the service was funded largely by Lord Suffield, who was seeking to create a direct route to market in order to transport grain and to profit from other land-owners who also needed to do so. Then, in the early 1960s, as Britain began to go on holiday, the Great Eastern Railway Company started running the 'Broadsman' and the 'Norfolk Coast Express', for city folk in search of the sea. By the late 1960s, though, when Englishmen discovered Europe, the service found itself chugging along on the chopping block, until it was christened the Bittern Line and was designated a community railway. Despite it cutting right through the Broads, the land on either side was drained centuries ago and even where it runs by the Bure Marshes, it would be an extraordinary event if you saw a bittern.

At Cromer, where tractors pulling balers are running up the fields, two sixty-something-year-olds holding hands – too in love for it to be their first time – get on and sit down across from me. 'Shire Horse Sanctuary – more like shite horse sanctuary' one of the boys across from me says and they all start laughing. Further along the line, the lady squeezes the man's arm and points out of the window to where row upon row of caravans are pulled up next to each other. 'Do you think that's a sales place?' she asks. 'Nah, that's storage,' he replies, as though he's been a student of caravans for many years. Two Union Jacks hang from the fence, running round the perimeter of the site, frayed and faded in the sun. She smiles at him and puts her head on his shoulder.

After Salhouse, a small voice behind me says we're almost there and a moment later she appears in the aisle. 'The train's going. The train's going, granny, and I'm even standing!' she squeals. 'Be careful,' comes the reply, and as we pass Carrow Road – the eighty-year-old home of Norwich City, built when the crowds grew too large for their old stadium, the Nest – the rest of the train start pulling their bags down from overhead.

Sweet cannabis smoke rises up from beneath the bridge across the Wensum and I peer over to see a couple, sitting on the top tubes of their bikes, passing a joint back and forth. The rain has stopped, boats are out on the river, and the warm hand of the sun runs across my neck. On Prince of Wales Road, Siciliano's wants a driver, Pure Gold is 'hiring dancers', and the barber shop is closed. 'CHAIRS FOR SALE' reads the sign in the window, a red felt-tip scrawl. Past the red-brick Agricultural Hall, built in 1882, which eventually became East Anglia Television's headquarters and was discovered to be built on top of an Anglo-Saxon burial site, I turn up into the heart of the city. The cafés are all too busy so I head on down the hill, cutting through university buildings where a grey-faced man, sitting on the remains of a medieval wall, is shouting into his phone. Out on Wensum Street a drain is blocked and the smell of sewage hangs over the bus lane. I walk until it fades and then I stop at a café where a big dog is lying across the threshold.

My sandwich takes half an hour to come and when it does the bread is flat and the grated cheese is turning back into a warm lump as though it's been in somebody's back pocket for most of the morning. Across from me a man opens his shopping bag and takes a record out to show his teenage daughter. 'It's 1986, Katie. The Smiths, 1986.' He smiles and nods when he says the date. For a moment she doesn't reply and he says it again, 'It's '86, Katie, when your mum and I got married.' The girl looks up, vacantly, from her milkshake – 'Cool,' she replies. Gently, he puts it back in his bag and asks if she'd like anything else.

On Magdalen Street, where wool merchants and bankers once came and went, delivery boys come and go now. Outside the pizza shops and chicken joints, they sit on their mopeds, parked in the gutter. Further up, sprawling Asian supermarkets butt up against Turkish restaurants and canine grooming parlours, 'second dog, a third off'. Where Magdalen Street runs round onto Waterloo Road, I check the address and start to think about what I'm going to say. I walk up and down each side twice, but the numbers jump from six to nine, and a Hawaiian-shirted man, holding a pair of scissors, pops his head out of the hairdresser's. 'Can I help?' he shouts across. 'I'm looking for Benny,' I reply. 'He's at number eight but I can't find an eight.' He looks at me, curiously, then turns his head over his left shoulder. 'The boy wants Benny,' he says to the two ladies cutting hair. They both look up in unison but I can't hear what they're saying. 'Sorry, love,' he shouts across. 'We don't know a Benny.' Thinking it's maybe a workshop out the back, I try to walk round behind the houses but there is no way through and I stand outside a flower shop on the corner wondering what to do. Behind the glass, a girl with a blonde bob in denim dungarees sits on a wicker chair, putting cactuses in small pots. I can feel her watching me, and when I look up she smiles. 'You're lost,' she says, when I walk into the shop. 'I think so,' I reply, 'I wanted eight Waterloo Road. I was going to see a taxidermist – if you know what that is – but there isn't an eight.' She stops a moment and tells me, shaking her head, that of course she knows what a taxidermist is. 'Show me the postcode.' I hold my phone down to her and she reads it twice, all the while potting more cactuses. 'That's out of town,' she says, looking up at me, 'you'd have to get a bus or a taxi. You can call one and wait if you like.' For fifteen minutes, I walk around the shop, thinking I should buy something, but by the time the taxi arrives, I still haven't decided between a Venus flytrap or lucky bamboo and I rush out the door.

Out of town, the wind blows through ferns on the side of the road and the brambles are ripening in the late August sun. 'You live in Hainford?' the taxi driver asks. 'I've never been,' I reply, then I start telling him all about the confusion and the girl in the flower shop. He glances at me in the rearview mirror and turns the radio up. 'Another absolute classic on the way,' the presenter promises, 'and extraordinary to believe it's 22 years since we first heard it.' The opening bars of ATC's Eurodance hit, 'Around the World (La La La La La)' fade in. The driver turns it up some more and we don't speak again until I get out and tell him to keep the 75 pence change.

When I turn left off the road, rooks take flight from the fir trees, crying out in alarm and swirling on the autumn wind. Halfway down the gravel drive there is a garage with an open door. I call twice but no response comes. The lights in the cottage all appear to be off but I can hear the muffled sound of a television and shadows flicker on a grey wall. There is no bell so I knock hard. For a moment, the TV stops playing and I take a few steps back, but then it starts again and there is no answer. All around the garden there are carved wooden animals and an old rusted truck sits on the gravel. I'm about to knock again when the thin figure of a woman appears from the side of the cottage. 'What do you want?' she asks. I tell her I've come to see Benny. I thought maybe he'd be in the shed. The woman shakes her head. 'Why do you want him? He's not here.' A magpie drifts overhead. She glances round to watch it settle in a tree then turns back and looks me flush in the eye. 'I was going to come and see him. It was just about the two bitterns.' She breathes in sharply through her nostrils and I start wondering what else I can say when she points to the end of the cottage. 'Go and wait there.'

For a couple of minutes I stand by the doors, and as my eyes start to adjust I can make out, through the glass, the outline of stuffed hares running along the back wall. When an internal door is opened the room is partially illuminated and a man with thick white hair shuffles across the carpet to let me in. As he walks back to stand behind the wooden counter, he gestures to the cased bitterns sitting next to each other on the floor. The larger of them, some 28 inches tall, plump mottled body and deep black pupils set in bright yellow irises, perches on a rock among plastic reeds. The smaller bird, with large green feet that it would have used to stir up water in search of prey, is set on a bed of mud. Benny leans on the wall, watching me in silence as I study the birds in the gloom, then he disappears back into the cottage, leaving the connecting door open. The telly is still playing and a wavering harmonica sounds, then three shots ring out, followed by the thunder of horses' hooves. When he returns he switches a light on and rolls himself a cigarette. 'That's something you won't never get again,' he says, looking down at the bitterns, 'that's Victorian.' He taps the cigarette on the countertop three times then strikes a match to light it. 'There's more about now than people think but they're secretive, you know.'

They haven't sold the type of tobacco that Benny is smoking for some years and a sour smell drifts across the room. 'That'll be male,' he says, pointing to the one on the left, 'clearer markings,

254

see, and the female is smaller.' When I ask him if he thinks of taxidermy as art he shrugs and tells me he's not bothered so long as he sells it, then he sits and smokes in silence. 'Nah,' he says after a while, 'it is an art form, it's got to be. That's probably more skilful and more art than a lot of things what people say are art.' He looks down at his cigarette then smokes the rest of it in one deep draw. 'Thing is,' he says, dropping the butt in an ashtray on the counter, 'the people that made them, they would have known exactly how a bittern moves. Same as it is today. Good taxidermy, you got to know exactly what the right positions is, course you do.' Benny says that he's never had a problem, not in five decades, but where it goes wrong for most taxidermists is the fat. 'I never handled a fresh bittern of course, but they can have a lot of fat and fat is always fat.' He stops and looks at me – as though wondering if he was right all along – then continues. 'Meat will dry, you see, but fat goes rancid.' Benny reaches for his tobacco, rolls another, then points to the door with a look of disgust. 'Some of these taxidermists, they're a joke.' He tilts his head back then blows a plume of smoke in the air. 'All that stupid stuff, people putting clothes on birds. That's not what I am about. That's no respect for the animal. You'd never see me do that.' Benny stops to pick a piece of tobacco off his tongue, then adds, 'But that's what sells these days though. It ain't right.'

Although he makes most of his income from commissions, Benny has found you can always sell a cased bittern. 'For a good one,' he says, like the big male, 'six hundred and fifty if the right person calls. Other one, about three hundred. Not a lot for what they are, when you think on it. They're antiques and they're a stunning bird.' Benny is in the middle of telling me that he thinks there are few people who appreciate creatures quite like taxidermists do, when the lady who told me he wasn't in appears at the door. 'Just popping out,' she says. 'Okay, girl,' he replies, 'see you later.' When she leaves, he walks round to stand next to me and looks up at a row of ducks. 'They're some of the prettiest things there are,' he says, 'just beautiful. Trouble is if they go out uncased people mess them about, start touching them. Whereas those bitterns, properly cased, are just as they were.' I crouch down to have another look at the smaller one and Benny grunts in disappointment, 'That's 120 years old. It ain't very good. That's just what it is. You know, just what it is.' I run my finger along the side of the glass and tell him I can see where the feathers are all out of place on the throat. 'They can be like that actually,' he replies. 'It's like people, we can all be different.'

Through a door with a coot and pheasant hanging above it, I can see a large stuffed fox and I ask Benny if we can go and have a look. He turns, without saying anything, and walks through. The creature is climbing a log with its tail swishing round behind it. 'Big, isn't he?' Benny says admiringly. 'Quite a handsome boy.' Benny shrugs when I ask him what he makes of people who don't like taxidermy. 'That's up to them,' he replies. 'That's their opinion. Nothing wrong with that, but it ain't mine. There's more people into it than you might think.' All around the room bundles of wood wool are tied, ready to be the basis of Benny's next projects. It's just the same technique, he explains, as the taxidermists who made those bitterns would have used, 'handmade forms and that. Otherwise you buy standard forms'. He looks at me and smiles, 'That's stupid though. Animals aren't standard. That's like putting your skin on me. You're a big guy and I'd have a lot of loose hanging down.' He walks over to a table with a set of scalpels laid out on it and picks up a brown bundle of wool and sawdust. 'Now that's not as easy as you like to think,' he says, passing it to me, 'because it's got to be the right sort of tightness to hold wires, that go up the legs and down the wings.' He takes it back and then hands me a bigger one.

'This would be just a bit smaller than the bittern, but it's the exact same and every bird takes about fifteen hours.'

I put the wool back on the table and Benny wanders over to the window. 'I'll tell you something,' he says, looking out. 'I encourage the small birds all about here. I leave my garden rough for the wildlife.' Outside, swallows are gathered on the telephone wire running across the garden. 'All them fucking houses down there got cats. All you do is see cats coming and killing my small bird population. It's about time somebody done something about it.' Benny puts on a posh voice, 'Oh, my cat brought me a present. He left it on the doorstep.' He shakes his head. 'Heard it so many times. That's my young birds. That just piss me off.'

On the way out, I have one last look at the bittern. 'Do you know how they boom, Benny?' I ask. He glances down at the birds, looking from the bigger one to the smaller, then tells me that, 'matter of fact', he doesn't. 'That's like anything. You know, how do they do it, that's like any bird you know, how do it sing? I don't know how they boom. It must be a thing in the throat, it must be. It's got to be.' I tell him I'll give him two hundred for the small one. 'That'd be no good,' he replies, 'it's got to be three. I ain't bothered one way or the other. I like having it anyway.' I'm almost at the gate when he calls, 'Do look after yourself', and when I turn he nods then closes the door.

A herring gull perches on a lichen-covered mooring pole, pale yellow eyes looking down along the quay. Elbow to elbow, children in shorts and fleeces dangle crabbing lines in the foamy brown water. A small girl turns and shouts, as loudly as she can, at a little boy standing 3 yards behind her. 'Do you want to look? I got one!' Twisting the orange reel, the girl drags the crab up the concrete wall before pulling the hook out from between its claw. She gives it to the boy who holds it right up to his face and then pokes it with his finger. The child jerks his hand back and calls across to a couple sitting at a table outside the café. 'Why does he bite me?' The woman, heavily pregnant, breaks off from her conversation to tell the toddler that the crab is a beastie. 'When Lola comes I'm going to catch loads more,' the little girl says, after she's thrown the creature back into the water. 'Really?' the boy replies in delight. She nods sternly. 'Yes, and then we will show you.' Wailing a strangled cry, the gull lifts into the wind before turning and drifting towards the sea, over sedge fading yellow beneath a heavy orange sun.

Up on the main street outside Pimpernel Cottage, two Land Rovers, one new and the other immaculately restored, face off. There's no room for passing and holidaymakers are spread out across the road. 'Quite a shock when she found out the truth about what he'd been up to,' a man in tight trousers says to the person walking next to him. 'Just goes to show you can never really know a fellow,' he replies, and they both smile.

At the top of the road, where the bins are overflowing and all the bags have burst, I cut up through the Pastures. On the grass, three families are having a picnic and two squirrels are running round and round, chasing each other beneath a beech tree. I walk on and emerge out onto New Road. Beneath the Blakeney war memorial lie last year's wreaths, pink plastic petals turning white in the sun. I stop for a moment to count the names, 31 men who didn't make it back to Norfolk, then I cross the road to Andy Randell's cottage. I want two crabs and to know if Henry managed to get onto the reed beds in the end. There is no car parked up, the lights are all off in the windows, and the shutters have been pulled across on the shed. Down the track, pigeon eggs from birds nesting in the ivy above have fallen onto the earth, white broken shells. In the first shed old creels and buoys are stacked up, in the second there is a pile of ropes, and then right at the end, on a sign leaning against a post, 'closed' has been written in black capital letters. When I'm back out on the road I notice that a builder on the other side, the face of a large dog tattooed on his chest, has been watching me. Above him, the sun is starting to descend behind the pine frame of a new house.

Down on the quay, most of the children have gone and I watch the last of them while standing in the queue for the shellfish van in the car park. Twenty yards away, a woman in her fifties helps her father up off a bench. 'Are we going to go and sit where we sat before, dad, to look at the ducks?' she says to him as she takes his arm. He doesn't reply and they wander slowly across the mud. They've run out of crab by the time I get to the front, so I ask the lady behind the counter for a pot of clams instead. 'Been a while since I saw him,' she says when I ask if Andy Randell is around. 'Goes away to Scotland in the summer, you see, gets too busy for him here.' She shrugs as though to say it is what it is, then calls out for the 'next customer please', as she drops my change into my hand.

Out northwest

Imagine them not here.
No sparrows chattering among the creeves.
No shrieking pickies, swooping down like knives,

The Island stripped of its glitter; no raucous wake
Towed by the trawler fleet; gone, like the corncrake,
The cormorants swept up from the blackened pier.

Katrina Porteous, 'Birds', 2019,

words that are to be written on a sculpture by the sea

The old lady digging in her garden with a fork, down at the bottom of the track, had said it was too early yet. The coldest April she's known and 'they'll not come in on these easterlies', but somewhere among the primroses, beneath the crow-stepped walls, a corncrake is calling.

Crooked and rusted red, a fence runs alongside the dusty drive and up through the empty window frames of the once-great house, the sky in the west is still lit blue by the last of the night-time sun. Seven hours earlier, after hurrying across the 2-mile sands that separate North Uist from the tidal island of Vallay, I climbed up over the pile of smashed slate and rotting plaster just behind the front door. Every winter the wind blowing straight in off the Atlantic rips out roof timbers and the pile in the hall grows higher, blocking off more and more of the rooms. They must have known I was there but they didn't take flight until I was directly beneath them, wheeling round close above the chimney stacks and screaming on the breeze. When I looked up I saw, just above me, their flightless raven chick perched on all that remains of the drawing-room floor, blue eyes bright with fear.

By the time George Beveridge inherited Vallay, the profits from his father's linen business, which had financed the construction of the mansion in 1902, had dried up. George could no longer afford to pay his servants and as the damp rose around him, he started closing doors until he had retreated into just one room, surrounded by the taxidermy birds that the Beveridges had shot on the island over the years. On 16 November 1944, George's drowned body was found washed up on the rocks just below me. Two days earlier, in the dark at high tide, he had set off across the strand in a boat.

For half an hour, I lie with my face in the bishop's weed and buttercups, listening out for the corncrake to call again, but it doesn't, and at midnight I pull my sleeping bag up over my head and curl in against the cold until the sun rises over the Minch.

When I last visited Edinburgh there was a piano bar on one side and an Italian doing takeaway pizza across the street, but in the early nineteenth century the opening at the northwest corner of Charlotte Square, 283 miles southeast of Vallay, ran straight out onto hay meadows. In his 1856 book, *Memorials of His Time*, Henry Cockburn, a shining figure in Scottish literary society, recalls the nights in which he stood at that opening and 'listened to the ceaseless rural corn-craiks, nestling happily in the dewy grass'. What Cockburn heard was soon to be lost, because by the time his book was published, the land those birds lived on had been planted with great Georgian townhouses. A century and a half later, when fewer than a thousand male birds return to Britain each spring – almost all of them bound for the periphery on the northwest – the idea that they once lived just beyond our cities is unimaginable, but when Cockburn was writing, corncrakes bred in every county, from Sussex to Caithness.

While building inevitably played some part in the destruction of their habitat, it was the development of the horse-drawn mower in the 1850s that led to the first severe declines. Corncrakes generally nest in tall grass such as hay fields, but labourers swinging scythes moved slowly, allowing birds to get away. Because the going was steady, the hay harvest took months and some chicks would have fledged by the time their patch was cut, but a horse could do the work of twelve men, meaning cutting was more likely to happen when corncrakes were still on the nest. Almost a century after horses pulling mowers put farmhands out of jobs, they were put out of their jobs, in turn, by the ever-improving tractor. In places such as the Pennines, by the 1970s the mechanised hay harvest took just half as long as it had done 20 years previously, progress that coincided exactly with corncrakes ceasing to breed there.

Back on North Uist at the Lochmaddy Hotel, tired, naked and out of the shower, I sprawl across the bed eating the custard creams that were left out on the tea tray. Just up the road there was a painter. She captured the terror of the land and the water in oils on big canvases and I wanted to meet her, but she wanted to be more than an artist of the islands and she left her boyfriend and her sheep.

'Just this last week,' the lady at reception told me when I arrived, 'her van's gone, the sign's gone, she's gone, no idea where.' At five, when I wake, the rain has blown in, my white belly is covered in biscuit crumbs and on a lamppost, beyond the window, a yellow sign reading 'VOTE SNP' flaps in the wind, held by a single zip-tie.

For two days it rains and I don't hear from Jamie. The Romanian builders down in the bar tell me they've only got three months left – 'three more months and then we go to Falklands'. Some of them have been before and they tell the others it's much like Uist, except for the penguins. On the third day, the lady behind the desk says it's Jamie's daughter who cooks the breakfast and she can take a message to the kitchen for me if I like. 'Down in the shed on the machair,' comes the reply with the eggs, 'but he doesn't work Mondays.'

Beneath my boots, dunes roll down to the sea, and behind me long narrow crofts run up from their yards full of rusted cars and sheep trailers, to grazing, to bog, to marram grass, to daisies, buttercups, and birdsfoot trefoil, all waiting to burst yellow when the sun finally comes. It's as though the waves are pushing on one side and the land on the other, creating a ridge that draws out above the beach then curls round and snakes away into the water. Five hundred yards inland, a woman in blue overalls with a dog at her feet is unloading silage for her cattle, and one field over, a newly built shed stands just above the high-tide mark, a tractor pulled up beside it.

Jamie Boyle is a small man and after walking round the shed, following the sound of a hot drill bit whining through metal, I can only see his head sticking up above the deep tractor bucket. I shout twice and the second time he lets the drill run out and peers over, the RSPB's most westerly employee. His cheeks have been burned red by the wind and he has a small hat with flaps pulled tightly down over his ears. He nods when I tell him I've been trying to get in touch, then he clambers down, in his ripped jeans, and leans on a sandy wheel.

It's been over 30 years since Jamie first arrived on North Uist, and although he still sounds like he's from Tyneside, he speaks with the cadence of the Western Isles. In the early days when he first came, back before he met his wife, some of the old ladies still collected birds' eggs for the table. 'There was one really nice lady, a bit eccentric, she went out and always looked for the first brood of lapwing eggs.' Jamie smiles when he tells me has no idea whether she knew it was totally illegal, but he felt it was best just to leave her to it. In his mind, that's about as far as it goes between crofters and wildlife. 'It's a funny thing,' he says, shaking his head, 'most of the crofters know which birds you can eat, they know the noisy ones, and the brightly coloured ones, but that's about it.' Every once in a while, meetings are called to talk about what more can be done for corncrakes, and Jamie tells me 'somebody always stands up and says just ask the crofters what's best. It always makes me laugh. I've never met a crofter who does know a lot about wildlife. I've just not come across that at all.'

Back and forth, in the shed, starlings are on the wing, their sweet voices sharp echoes against the cold aluminium walls. For all that Jamie thinks crofters don't know much about birds, he tells me that their survival is vital for the survival of Britain's corncrakes. Uist currently sustains about 300 calling males, most of them on the machair. As he speaks, he turns to face inland towards the sandy fields and dunes. 'Without this mosaic you wouldn't have the corncrakes here. What we've got is this grassland and crofters have all got parcels. Across two-thirds of the machair they take the sheep and cattle off on about the fifteenth of May, giving it all a break and giving the corncrakes somewhere to nest.' Back in the 1970s and '80s, the corncrake population on Uist was about half of what it is now, but it started to be recognised that if crofters could be encouraged to mow later, chicks would stand a better chance of fledging. 'The science,' Jamie says, nodding in approval. 'There's been a lot of science. A lot of work's been done. People like Tim Stowe and James Cadbury were the first. They really counted the corncrakes.' Cadbury, a member of that dynasty of chocolate makers and ornithologists, concluded his landmark 1980 study by saying that the best chance corncrakes have got is 'the crofting lands of the Western Isles, with their small fields, which are not cut until late July'.

While late July is much better than taking a first cut in May or June, as is often possible in southern England, corncrakes are generally double brooded, and in Jamie's experience the second lot of chicks normally hatch right in the middle of July, meaning they're vulnerable. Out over the water in front of us, a large white bird is drifting through the mid-morning gloom and Jamie steps down off the concrete base the shed sits on and stands among the nettles. 'Is that an egret?' he says, looking up at it. 'I think it is, you know.' He watches for a while as it cuts away, low over piles of crabbing pots, then high over the houses, before he returns to lean against the wheel. 'So anyway,' he continues, 'corncrakes have become really reliant on the late cut system. The first option is to cut on the first of August, then midway through August, and then the first of September. It's basically compensation. The grass is at its richest in July and would really make the best hay or silage then, so the crofters are compensated for a loss of quality.' Jamie shrugs when I ask how much they're generally ending up with and says it's maybe 'just a couple of thousand pounds kind of thing. It starts at £230 per hectare and goes up to about £400.' In years gone by, it was more, but the government now seems to be pulling back from the schemes. 'It's absolutely crazy,' Jamie tells me. 'They've done fantastically well building numbers up and now they're talking about taking the money away.' Over the fence, the lady who was feeding her cattle is driving a tractor up and down the fields. Her dog is sitting at her shoulder in the cabin and three gulls hang up above her on the breeze.

Jamie looks at me occasionally as though he thinks he should, but he seems more comfortable looking elsewhere. 'People talk about the machair,' he says, while watching the tide running in, 'and it translates from the Gaelic as fertile plain, but it's really a patchwork of things. You've got dune, then it comes down to dry machair, then sort of wet machair, then freshwater loch and then it goes into fen, and you've got these sort of fallow bits. It's seven or eight fantastic habitats in a really tight area right down the island.' Over the past 30 years, Jamie has seen crofting change. 'There's a real dearth of young crofters. The majority are quite elderly and you know, keeping cattle here is so marginal. There's just about no crofters now who have sort of three or four cows. There were some people only had two but they've all gone.' A mile or so out, over the Sound of Harris, the sky is clearing and the sun is burning through. 'Without the crofters,' Jamie says, turning to look up at the fields, 'it would eventually just go into ragwort and rank duney grassland.' He pauses a moment and then pulls a bolt from his pocket and starts rolling it back and forth between his finger and thumb. 'You'd lose your corncrakes,' he continues, 'you'd lose everything in fact.' I tell him I'll let him get on and he nods, but as I turn to go he smiles and points to the shoreline. 'My family live in a fantastic house there. You could see right across to the Sound of Harris but my father-in-law, he was a crofter, he just built an enormous shed at the back of the house.

He was so proud of it and it blocked all this off. You just think why, why would you do that?'

Angus Kyles cups his right hand in front of me and rests his foot on the rail. 'As I say, I've been over in Germany judging four times. The calves are beautiful but they seem to have forgotten about the udders.' He shakes his head in disbelief, holds his hand up higher, and curls his bottom lip over his teeth. 'A teat needs to be the right size, you see, so a calf can suckle on a cold day.' Angus leans over the rail and runs his hand across the dark red rump of the Highland bull. 'This fellow here is probably my best one. His brother made eleven thousand at Oban two years ago, so we'll be hoping to get similar for him.' The animal dips his big head round slowly, eyes us for a while, then goes back to eating. 'When I was crofting, as it was in the beginning, in 1986,' Angus continues, 'I only had twelve cows, and between lightning strikes and getting caught up in fences and drowning, four of the best were dead by the end of summer.' He turns away from his cattle and leans on the rail. He is a short man, red haired, built like a little bull, and he stares up at me. 'It was a disaster. That, I thought, was it. I thought I'll give up, sell the rest, and go to work on the mainland, but I hung on in there and this is me now.' The air inside the shed is a warm musk of cattle and oil and grain. Angus laughs when he notices me looking about. 'I'll tidy it up yet,' he says. 'I'll give it a clean yet, but it's been a hellish busy spring. I can see us feeding until the second week of June. Not the whole 160-strong herd but certainly a good chunk of it.'

Angus walks through the shed and sits on the edge of a large white tub that reads 'salmon' on the side. 'It was my mother's croft,' he tells me, pulling the sleeves of his green overalls up over his freckled arms. 'When I inherited it, it was 40 acres, but then I've been basically building it up ever since. Just working away. Shrewd investing here and there.' He holds his hand out and manoeuvres it through the air, like a fish swimming upstream against the current. 'Whatever goes on, you know. Working non-stop.' When I ask Angus who owns Vallay, he smiles as though I've asked the right question. 'I bought it all,' he replies brightly. 'I've got nearly 5,000 acres of ground now. It's bought and nearly all paid for. Except for the mansion. That belongs to Granville, North Uist estate. He owns most of the crofts.'

It is dark at both ends of the shed and birds flit from one end to the other, seemingly catching Angus's eye as they appear briefly in the light, before disappearing. 'There wasn't a gate over on Vallay when I started. There was nothing. It hadn't been cultivated in 60 years. There were no corncrakes, and then when we started cultivating and leaving it in the summer, that's when the corncrakes came along.' The Macaulay family, who sold Vallay to Angus, used to graze the island all summer instead of all winter but he couldn't understand why. 'It just makes no sense. Highland cattle like the hill. They want to be up on the hill ground. They want onto the hill in June.' He gestures over his shoulder and glances round at the shed wall. 'They want the minerals in the hill. They're pining for it, basically.' When Angus was a boy he remembers the hill ground being covered in thousands and thousands of cattle, all of which had been taken off the machair, but he thinks that increasingly crofters are using the wrong stock and he would like to see Jamie Boyle and the RSPB encouraging them to go back to the old ways with some sort of subsidy. 'The whole idea is that you clear the cows off all the ground, let it flourish, let it grow naturally. Everything should be on the hill, but you need stuff that will survive, Galloway cattle, Highlanders, Luings, things that are powerful in the hill.'

Folding his arms across his chest, Angus tells me that would be his way of doing it anyway, and 'then the corncrakes would flourish in the grass with nothing to stop them'. For a moment he stops talking, leans back, nods in thought, then begins again. 'I tell you another thing, you don't see many bees now. There used to be bees in the thousands. You know, you used to be turning the hay by hand and you could guarantee, after a few days, there'd be nests of bees in every row, but we don't have that anymore.' Outside the shed, the sky has darkened and rain is starting to fall above us, coming down noisily on the corrugated plastic skylight.

'We exactly do,' Angus replies when I ask him about Jamie telling me that the crofters don't know much about birds. 'That's the main feature why I do so much of what I do,' he says, shaking his head. 'I want wildlife there and the plants and the birds and everything else that runs around.' He stops and scuttles his fingers along the edge of the white container as though his hand is some sort of little creature. 'This would be a very dead and dull place without corn-crakes. What's the point if you don't have any wildlife? You know, the starlings and things like that coming in here. They come to steal my grain and they're very welcome to it.' He pauses and looks up, but the birds have stopped flying, and then down at the other end of the shed, in the far corner, a heifer with a calf starts lowing. 'She had a wee bit of a problem, you know, but she'll be all right soon.'

One of the biggest difficulties, as Angus sees it, is that people find it hard to make any money from a croft. He pushes his hands into his pockets and leans towards me. 'Very little return, but it can all be made profitable. When I was at school I was spreading seaweed on the croft to fertilise it, you know, and now I spread thousands of tonnes of seaweed a year.' I smile and say something about him having made it all work. 'Directly,' he replies, 'I've made a fortune. I've bought the land out of it. I didn't inherit that land. I bought it all and I've still made a constant profit every single year. I've never made a year loss yet on the farming side.' Angus reckons his mother, Ena, thinks he's off his head. 'Can't understand why I do it or how I do it'. He laughs and suddenly loses 20 years. 'She says why don't you just stick with 40 acres and be content with that? I know well she often says to herself, why is he doing that? She says it to me too, and I just say because I get a buzz from it. It's a thrill.'

Angus Kyles isn't his real name and he wasn't born on North Uist. He was christened Angus Macdonald but there are so many Angus Macdonalds on the Western Isles that many of them are referred to by their business or whatever their croft is called. 'They divorced when I was two, my father and her,' he tells me. 'They were in Australia. He stayed and I came back with my mother.' Angus says he could never understand where his drive came from and why he wanted so much. 'I didn't know my father till I was 32 and then I went to Australia to meet him. I sat down for a few hours and we just talked and I realised then why I am the way that I am. I realised then where I got it from.' Just as I'm asking what his father did, he cuts across me. 'You know, it's funny, genetics are a strange thing. I think about it a lot and it's definitely the same in cattle and the same in sheep, definitely. My father, he died not long after that.'

Angus stands and walks over to the red David Brown tractor in the near corner. It is over five decades old but it looks perfect. Three years ago, just after she finished university, Angus's daughter died and her mother decided that they should go on a one-million-mile journey in her memory. Part of it was completed by Angus circumnavigating the island on the old David Brown with 60 other crofters, following on behind. 'I was 10 years old when I started driving this,' he says, drumming his fingers on the red bonnet. 'My mother would always be on the reaper, you see, cutting the hay.' Until recently,

Ena was still crofting, but after having a stroke 12 months ago she now lives quietly. 'Gets a bit stressed and anxious,' Angus tells me. 'She can barely get her hens in and out. It'll happen to us all, but it's hard going from being a person who was so very fit to being a person who can't do a thing overnight.' It takes him a bit of time to work it out when I ask, but he tells me that Ena would have been three years old when George Beveridge was found dead on the rocks on Vallay. 'Her brother used to work over there actually, but I can't believe what happened or why it happened. That family were so wealthy, there was no way in hell they should have went bust.' He looks up at me and shakes his head. 'It does make you wonder,' I reply, 'whether he did just drown.' For a moment he sucks at his lip and I regret saying it at all. 'You just don't know,' he replies, eventually. 'They were too big into the one thing, and when you've had generations of wealth, it's extremely difficult to cope with adversity. I'm a great believer, Patrick, that nothing comes to you. You've got to remember that and remember too that competition is the healthiest thing in the book.'

He tells me he's got to get back to work, then zips his overalls up and turns to have one last look at his cattle. Out in the yard a big white tractor, just delivered from Finland three days ago, glistens in the rain. 'Good luck,' Angus says, climbing up into the cabin, then he sits and makes a call. 'The whole world has just come to a standstill,' he says down the phone. 'That's me been waiting three weeks for it now.' A part is late and he needs to get on.

All down the A1, pigeons in the gutter gather grit. The combines are rolling and they need it in their crops to grind the grain. At Rosebrough, I pull over onto the side of the road and push my way through the ripening elder to piss out of sight of the truckers going south. It's only 9 miles to Beadnell, but the world isn't ready yet it seems for strangers to start peeing again, in each other's houses.

In the photograph, Katrina Porteous, camouflage bucket hat and blue corduroys, sits on a bench between two old fishermen. On her right, Charlie Douglas, with a bicycle clip around his trouser leg, looks away from the camera, and to her left, Charlie's brother, Tom, smiles broadly from beneath his woollen cap, a cigarette held between bony fingers. Katrina, at that point, was only 30, but she was already well known to what remained of Beadnell's fishing community, as the young poet who hung around the huts and wanted to know everything.

When I arrive at the cottage, all the lights are off and I walk round to the side, across the grass, trying to work out if I've come to the right place. It looks both very lived in and like nobody's lived there for 30 years. Beneath the window, there is a pile of books on a faded turquoise carpet, and when I press my nose to the thin glass, I can make out a row of china plates hanging on the wall above a tiled fireplace that was probably white once. Back at the door I knock twice, then I turn round and lean on the wall. Down on the beach, over the road, two small boys are running nets through a rock pool, shouting about shrimps, and in the thick late summer air the sweet rot of the sea hangs heavy. As I raise my hand to knock a third time, footsteps sound, a bolt is pulled across, and Katrina appears, crooked teeth, bright blue eyes, a jumper covered in stars, and smiling the very same smile as when she sat between the Douglas brothers three decades ago. For a moment, she looks down towards the beach where the boys are playing, then shows me to the kitchen, weaving her way through the houseplants in the hall.

Her mother, Katrina tells me, never really thought much of her moving into her grandparents' cottage, in her twenties, to write about the fishing community. 'She was from a Durham pit village and it was something you didn't do because you wanted to better yourself. I think for a lot of people that means moving to the city, but really this has sort of been my life's work.' Katrina pulls a face like a happily defiant child, then puts the kettle on the hob, and wanders to the kitchen window. 'Until 15 years ago,' she says, looking out, 'this was pasture and you could see all the way to the border, right to the Cheviots, and I remember the corncrakes as a child.' Twenty yards out, just beyond the fuchsia bushes, two houses with solar panels on their roofs loom over the garden. Katrina spoons coffee into a pot and places it on the ironing board next to a vase of wilting roses sitting on a pile of yesterday's papers. 'Not many of them are lived in,' she says as she takes the kettle off the stove. 'They're owned as holiday lets and the visitors have no attachment to the place.'

In Beadnell now, there is just one fisherman left who still lives in the village, but back in the 1990s there were always men mending creels and the harbour was full of boats. 'It's why I loved living here for the first 10 years,' Katrina tells me. 'There really was a community and I came to care about it deeply. I cared about its values and its future as well as its past, and its place among the rocks and the birds.' She sits and leans forward in her chair, then says, scrunching her eyes almost closed, that along with the rocks and the birds it gave her an understanding of 'big time'. Katrina gets up, puts four white rolls on a tray and places them in the oven. 'Their relationship to corncrakes,' she says, as she roots around in a cupboard above the sink, 'was very close. They were important. They remembered them and they missed them.'

From conversations she's had over the years, Katrina thinks it's absolutely true that for fishermen born before World War I, they almost didn't think in terms of dates at all. Instead, they knew it was time to do things like paint the boats again because the corncrakes were calling in the night and the swallows had come back. Three generations later, though, rather than being a bird that signified continuity, the corncrake, in its absence, came to symbolise great change. 'They talked about them a lot,' Katrina tells me, 'in the sense that they were disappearing. It was a measure to them. Tom, who you saw in that photograph, would say to me that the fields were full of corncrakes. Just like football rattles, he always said, and he'd hear them when he was on his bike early in the morning.' Katrina takes a plate from the sideboard, next to where a heap of cold tea bags is piled up on a saucer. 'I'm afraid I don't have much in but we'll have these rolls and there might be some cheese.'

Katrina shrugs slightly when I ask her if the fishermen knew why the corncrakes had gone. 'Just greed,' she replies. 'Charlie would have said it was greed in the end. They filled those fields with houses, far more than had been originally planned, and the locals were priced out.' As Katrina is trying to open a seized lid on a pot of raspberry jam, the phone rings and she looks up at the clock then tells me she has to answer it. Her father is in hospital. When she goes, I'm suddenly aware that the old Prestcold fridge in the corner is humming to itself softly and making occasional sickly gurgling sounds. 'I can't speak,' I hear Katrina saying in the other room. 'I'm sorry, I can't. I'm with someone.' By the time she returns, I've managed to get the lid off and she takes it from me and smiles anxiously.

When she sits, she pours me a cup of coffee, then tears at a roll, eating it piece by piece, spreading jam on each individual chunk. 'You know, I can't remember vast numbers of corncrakes,' she says between mouthfuls, 'but I do remember being with my grandad.

275

It would have been 1966 or maybe '67 and he would point them out. It was that sound, and sound has always been incredibly important to me.' Katrina believes poetry starts in listening and she feels that, to truly understand a place, it's essential to pay close attention to the aural landscape. 'I think sometimes that a recording of a place, a sound recording, is more evocative than a photograph and birds are the animating spirits of place, and that I would say is the role they play in my work.' In the upstairs window of one of the houses out the back, a topless man with a soft pale belly lifts a large box onto a bed and pulls something down across it, a screwdriver or a blade, then opens it up and looks inside. I watch him over Katrina's shoulder and she looks at me strangely, running her fingers through her black hair. 'But I also think,' she says, spilling a few drops of coffee on the tablecloth as she fills up my cup, 'that birds express things about being human. They are deeply embedded metaphors about our longing, our longing to escape, and our gravity and physicality.' As she speaks, she closes her eyes again and her mouth tightens in deep thought. 'Because they can fly,' I reply, 'and we can't.' She opens her eyes again and smiles. 'Yes, in part, because they can fly.' I lean back in my chair, yawning and stretching my hands out behind me and then they touch warm fabric and I lurch forwards. I'm not sure what I thought it was or why it frightened me. Katrina laughs and brings her mug down on the table. 'Sorry,' she says, as though she really means it. 'My electricals are very weird in this house.'

Bald and with cheeks covered in a web of burst blood vessels, the ice-cream man leans out of the hatch to talk to the two little girls below. The smaller one glances up at him then whispers, anxiously, into the bigger one's ear. 'My sister wants hers in a tub,' the older girl tells the man, before pulling a £2 coin out of her pocket and stretching up to place it on the counter.

Katrina and I are standing together on the harbour wall looking out over the grey sea towards Dunstanburgh Castle. During the early herring years, there would have been as many as a hundred boats tied up beneath the orange and green lichen-covered stone, but the fishing is hardly worth it now and there are only two boats left. 'I often think,' Katrina says, turning to look at me, 'that we get very excited about ancient ruins, but in birds we can hear the same sounds people were hearing 10,000 years ago. Those corncrakes I heard as a child. That's a sound that Bronze Age people would have known. I find that really reassuring.' The two little girls, one with her cone and one with her tub, skip past us in their pink sandals and then they sit, 20 yards away, swinging their legs on the wall. 'Birds are a reminder of our place in history,' Katrina continues. 'Birds have been here for longer than we have and somewhere they'll probably be here after we've gone. We may not be here very much longer.' Down along the harbour wall, ice cream smeared all across their faces and chocolate flakes held between chubby fingers, the two little girls stare up at us.

We turn and walk up past the old kilns where lime was burned, and then herring was salted, and where old broken lobster pots rot now. In the village, the pavement is full of dawdling holidaymakers, all moving towards the beach in a ragged, sunburned, screeching procession. Katrina steps into the road to go at her own pace and I follow. 'In a way,' she says, smiling, 'I've devoted my life to concentrating on what are depressing truths, but I increasingly don't find them depressing because if we continue destroying the earth at the rate we are, then the earth will get rid of us.' Up ahead, where a sandy track runs up to the dunes, a woman is shouting at her toddler. 'What do you want?' she says to him as he wriggles to get free, his hands pulling at the straps of his pushchair. The boy looks up at his mother and starts to wail. Next to him, panting and with a leg at each corner, a French bulldog is tangled up in its lead. 'I'm not uncaring about people,' Katrina says, as we walk off the road and head up the track. 'People matter to me deeply, I mean, more than I can sort of express really.' She walks on ahead of me and raises her voice so I can hear. 'At the same time though, humanity is just a glint on the surface of the sea. It's one little speck of light in the time that the universe has occupied.'

I look back to see that the mother has turned around and is taking her screaming boy home. Katrina stops for a moment to take her boots off and then turns towards the village. 'When Charlie and Tom were young men, they had incredibly hard lives,' she tells me. 'There were all these bonds and the birds were part of that, but at the same time, they slept on sails in the attic because there weren't enough beds. They went without shoes in summer. Seven brothers, and they went without shoes to save leather.' Katrina carries on 30 yards further up the track and stops at the top of the dunes. Below us on the white sand, families are setting up windbreaks and children paddle in the shallows. 'I wouldn't want to go back to that life,' she says, looking down at the beach. 'There was space for everything and the fishing was sustainable, but I wouldn't want to be a woman in that culture.' Katrina shrugs apologetically, when she tells me she's not a romantic. 'Not really. Capitalism has given us huge gifts. It's given us a great standard of living, but it's severed all those bonds.' She looks at me as if wondering whether she's saying things you really shouldn't say. 'I mean, I think we've got ameliorate it, but I think industrialisation is almost an organic process.' Three children behind us are waiting to pass and Katrina puts her hand on my shoulder to move me to the side while she carries on talking. 'It's almost a virus or a disease, the sort of cultural progress industrialisation represents. I don't think it's stoppable.' She walks on ahead of me along the top of the dunes then turns and smiles. 'This,' she says, pointing down to the caravan park, is what she really wanted to show me. 'It was about 10 years ago. I heard it and I thought it can't be. It was a sound I hadn't heard for a long time but it kept calling and calling.' Katrina pauses, then thrusts her hand up into the air. 'I watched and watched, then its head popped out of the grass. It was a corncrake.'

Down at the water's edge, 100 yards below us, a man is slumped on his side blowing up a green rubber inflatable. Every 20 seconds or so, he pauses, comes up for air and then starts again. Further along the beach, where children's voices fade beneath the rising screech of little terns, I glance back to see that the inflatable has become a crocodile and the man is 20 yards out drifting in the breeze.

Goose pimples on pale arms and a blue woollen hat rolled up over her ears, Cheryl McIntyre appears back at the kitchen door after throwing a bucket of scraps to the hens. 'I don't like the north wind,' she says, looking behind her as she slams it shut. 'It's a wee bit strange. It kind of blows the door open.' She keeps her boots on and treads dirt across the brown linoleum to the cooker. 'We could do with it being a bit drier for cutting,' she tells me, lifting the kettle off the hob, 'but we're meant to be getting rain on Sunday, so anyways, that's fine.'

Behind me, through a low window, a wooden boat lolls lonely on black rocks and the hard sea runs across the bay to Broadford, Skye's second-largest town. Cheryl moved up to the island just over a decade ago, and 'Rural Skills', at the University of the Highlands and Islands, was her third degree. She had already at that point

studied English at Glasgow before going on to do a teaching qual-
ification. 'I am a wee bit settler in a way,' she says with a smile as
she pours boiling water into my cup over two spoonfuls of coffee
granules, 'but at the same time there's a lot of old boys sitting on
crofts, not passing them on to anyone, and the crofts are dying.'
Cheryl goes to the fridge, opens it, and asks me for the second time
if I'm vegetarian. I shake my head and then she shakes hers. 'Sorry,
sorry, I asked you already. Right, I'll do bacon rolls.'

In the corner of the room, Struan, a large beige dog of uncertain
breeding, starts barking and then the door to the hall opens. 'Tractor's
fine,' Stephen says, as he walks to the far window, phrasing the
words as a question. 'Fine,' Cheryl replies cheerfully, 'the blades are
all out but there's a few wee tufts just not getting cut, but if I max
up the revs, it's better.' Stephen nods, then sits with his legs spread
and his hands pushed hard into his pockets. 'I hear you were up
fixing the tractor this morning?' I say to him. He looks up and
glances across at me. Small dark eyes and thick black hair, he must
be a couple of years younger than Cheryl and is good looking in
some vague way. 'It's all I do,' he replies.

As Cheryl cuts the rolls, she tells me it works well between
them, because they can fix anything that needs to be fixed while
also applying for every grant that comes along. 'I would say crofting
was almost never designed to make people a living, because it
meant they had to work for the estate.' Currently, almost every
crofter has a second and often third source of income, but as well
as grants to cut late for corncrakes, there are a number of other
schemes that can provide funds. The house we're in and the land
out the back aren't Cheryl's. The croft belongs to a man called
Angus McHattie who is away farming cattle in Sweden. She tells
me almost nothing about him except that 'on his bedside table,
there is a stack of books on wolves'. The deal was easy enough:

if Cheryl looked after Angus's cows and did his paperwork to get him into some of the schemes, she and Stephen had a place to live and could run their stock on his croft. Cheryl tells me she thinks that due to the applications being as complex as they are, roughly only 5 per cent of crofters on the Western Isles will have tapped into funding that incentivises them to manage their land in a way that benefits corncrake. 'I was out this morning,' she says, sitting down and passing me a roll, 'cutting silage on another croft along the road. It's not in a corncrake scheme but it should be. It's just that the paperwork is preventative, so the crofter who owns it isn't really interested.'

Just over a year ago, Cheryl got the news that her application for a tenancy on a croft of her own, 30 miles across the island at Portnalong, had been accepted. 'It had just been lying there for 20 years and nothing was happening. People were popping cows over to it now and then but it was derelict.' So far, she says between mouthfuls, they've had 'a Young Farmer's Startup Grant of fifty grand', and they've now raised the frame of the house they're building, but with everything else that's been going on, progress has been slow.

Without saying anything, Stephen walks across, puts another roll down in front of me, then sits back by the window, looking agitated. 'So you work as a mechanic?' I ask him. He nods as I tell him about an old car I used to have that leaked engine oil all over the road. 'Had a few 1980s cars myself,' he replies. 'I wouldn't do it again though. I'd happily drive a van for the rest of my life.' Cheryl looks round at him, smiling, and then gets up, goes to the fridge, and takes out a carton of orange juice. When she pulls on an old jumper and tells me it's time to crack on, Struan pads to the door and looks round at her expectantly. On her throat there is a shaved patch of fur, where she had a lump removed. 'Right,' Cheryl says, putting her plate down on the draining board by the sink. 'If you're done, we'll go and cut this field.'

As we walk up the garden towards a small gap in the hawthorn hedge, the dog lopes on ahead of us. 'It's a funny thing,' Cheryl begins, before pausing to take a gulp of juice. 'I've never actually heard a corncrake here on Skye. Who knows, but I don't think they've been here for 20 years. In a way, corncrake habitat is a big part of my life but the corncrakes themselves aren't.' Up the field, beyond the tractor and over the fence, the ground rises up in a large circle. They aren't visible from where we're standing, but on top of the mound there are two slabs, 8 feet apart, which are said to cover up the entrance to a Neolithic burial chamber.

Looking down along the other crofts, Cheryl tells me that something has changed. 'People have stopped cutting, but you know on Skye now we have sometimes three houses per croft, often holiday houses. Before, it was quite untouched, so maybe that's one of the reasons the corncrakes have gone.' Cheryl walks a little way further on, then pulls up a clump of meadowsweet and tears at the white flowers with her nails. 'We get basically a hundred and fifty pounds per hectare for leaving these rough refuges. It's another option called Management of Cover for Corncrakes. It's just a big drainage scar, but there's a mix of herbage that provides shelter all year.' The blooms of angelica and vetch are wilted, purple gone grey, and the dock leaves are peppered red, dying back beneath the fresh autumn sky. 'Quite a nice wee bit,' Cheryl says as she carries on to the tractor. She walks round it once slowly and then crouches down behind the mower. 'When it's wet you get frogs run over and quite a lot of wee mice get run over too. It's a wee bit Rabbie Burns.' Cheryl runs her fingers across the blades and tells me she has no option with this field because of drainage ditches other than starting on the outside. 'What I do though is I leave a section in the middle and I cut that at the end. It sounds like you're just working your way in, which is obviously the old evil cutting, but I'm leaving a channel.' She pauses and sweeps her hand through the air. 'The theory is that any young corncrakes will get pushed out the top.' She climbs in and turns the key with a doubtful look on her face, but the tractor shunts to life, a rich gurgle in its oily guts. 'Too much coffee,' she says, as she jumps out and runs over to the fence, before disappearing for a moment among the bracken.

At the margins of the field, rowan trees are heavy with scarlet berries and the last of the swallows chitter on the breeze, drifting impatiently around the telephone wires, settling and then taking flight again, waiting to fly back south when the wind changes. Cheryl wanders back, tucking her t-shirt into her jeans, then climbs up the metal steps into the cab. As she takes off up the field, a dark plume rises in the clear air above her and Struan runs behind in anticipation of the dead. After 20 minutes, Cheryl stops and leans out of the tractor. 'What I'm quite keen to do, given I've got the mower up here, is just get the second field done after I've finished here, but you've got the directions to my caravan.' She turns round and points to the hill across the bay. 'That's Glamaig, see, that hill there. It's the same distance again after that, but you've got the postcode. There's a German guy there as well but I've told him you're coming.' I tell her it's no problem, trying to remember if she ever mentioned a German guy, and then I ask again if she'll be at the croft in the next few days. 'I do have to get across there at some point,' she says. 'I'll make it over on Friday night.' She draws the door closed on the cab, pours the last of the juice into her mouth, lets the mower down, and then runs away up the field, cutting carefully for a bird that isn't there.

For the rest of the day I drive around the island in no particular direction. In spring, just 10 male corncrakes were counted calling across Skye, and at Waternish, up on the western edge, where three of them were heard, the farmer I wanted to see is out. By the Fairy Pools, an old couple have run off the road in a new motorhome and caravans are backed up. Over in Broadford, outside the bakers, an Englishman in a too-small t-shirt roars at his son. 'I didn't want a sausage roll. I wanted an actual sausage in a roll. I never have a sausage roll!'

Across at Portnalong, a coach with a mattress on the floor and a woodburner in the corner rusts away at the end of a track. On the side, beneath a sun-bleached saltire, black lettering reads STRATHPEFFER but it doesn't look like it's been anywhere for some time. Further down at the bottom of the croft, behind a shed, a beige caravan is pulled up beneath a willow tree on the edge of a wood that grows out of a cliff face falling away to the sea. For two hours I sit, watching buzzards out the window, and then at seven he arrives, sunburned, smiling, just gone 40 and holding a bag of lager. 'That old bus is where Stephen and Cheryl were living,' he tells me, as he sits down and opens a beer, 'but this is what Cheryl always wanted even when she was quite young and she lived in Glasgow with Rosslyn, who is now my wife.'

All evening we talk. It was going well for Kris Pohl back in the 1990s when he was playing bass for Schrottgrenze who were doing big things on the German powerpop scene. 'Things fall apart though,' he shrugs. 'It all started to go wrong for us, as a band, when we refused to support a kind of political movement. It was about Germany getting back to new strength. The name translates maybe as "new us". It was this kind of bullshit.' The guys who ran their record label were keen they backed it, but Kris tells me they just couldn't. In the end, he thinks it became everything they thought it would, but by then it was over. Without the support of the record company they were down to playing small gigs in towns they'd never heard of. Kris goes to the fridge and gets two more beers. A little over a decade after he left, Schrottgrenze's lead singer, Alex came out as gay and now often appears as Saskia Lavaux, a blonde-haired woman with heavily made-up eyes. Kris says it was a lovely thing because he worried for some time that Alex was unhappy and he feels it made life on tour unhappy. In part, their art was born of being angry with all sorts of things, but sometimes it was too much for him. 'So they started writing new music and they asked if I wanted to go back,' he tells me, but by then Kris was in Glasgow with his own unhappiness. He was mastering and restoring audio but desperately wanted out and had dreams of becoming a forest ranger.

Outside, a small sweep of starlings drifts as one, contorting and twisting high above the trees. Kris tells me that when it gets dark, he usually 'turns off the lights and simply has a candle, otherwise the midges will come'. A smile burns across his face when I explain I've come about corncrakes. 'There was a sign, old and kind of broken with water,' he says as he opens his third beer, 'when I went once to North Ronaldsay. It said to ring a number if you hear or see a corncrake.' He throws his head back and makes a long rasping noise in his throat. 'We heard them like that all the time. We were right in the middle of them and it wasn't until we returned home that I realised how lucky we got.'

Every day I write and every day Kris goes off to deal with problem campers: hundreds where they shouldn't be, three men trashing a wood spangled on ecstasy, and four fifteen-year-olds with their first beers. Then, when he comes back, we sit and talk. We talk of the future, of how proud he is of his wife who is doing a doctorate on Lilias Skene, a seventeenth-century Quaker poet, and we talk of the past. When Kris was a child, he tells me he would climb into his grandparents' bed and ask them for stories about growing up and what it was like when they were young. 'It was always dragons and knights. They always had stories about dragons and knights. They were such loving people.'

At night, Kris wakes and I can hear his piss hit the bowl. 'You hear everything in here,' he tells me one morning. 'When I peed I could hear you breathe.' On the Friday, I wake late and the day beyond the fogged-up window has begun. Cheryl sent a message at six: 'Hope had a good time, problem with silage, won't make croft.' I run my fingers over the bites on my pale belly and then walk outside and wander up a track that has been beaten away through the bracken. The foundations of Cheryl's house have been built out of grey brick, and the metal frame, some 20 feet high, has been painted red.

Shrouding the Cuillins in the distance, an autumn mist has come down. 'A rich man tried to sell them to fund the refurbishment of his castle,' Kris told me the night I arrived. 'A man selling a mountain.' He said it as though it was one of the funniest things he'd ever heard. I leave him a note and tuck it under the toaster, 'Good luck, my friend', then I drive to the bridge.

Outside the Tradewinds pub on the road to Fort William, a big red T sits at a jaunty angle above the door. Fifty miles south, after Tyndrum, I pass a river where we used to stop and swim on our way back to school every year after a trip to the Slate Islands. The teacher who took us seemed to me, at that point, to know all there was to know about birds. He told me everything and I forgot it all. I was in my final year of university when I heard he'd hanged himself. I'd been to see him at Christmas time, when I was home, and only a couple of weeks after that somebody said he'd done something a long time ago. Between Crianlarich and Callander, the traffic is heavy.

Camper-vans crawl and motorbikes rush by, the world going south as the weather turns. On the outskirts of Edinburgh, it's closing time at the zoo and a group of small school children, holding hands in twos, stand outside the entrance in tiger masks while the teacher takes their picture. A biting wind is blowing when I arrive in Charlotte Square and I sit for a moment at the entrance to the northwest passage where the corncrakes once called. Across the street to my left, a bus driver perches on the steps outside the National Records Office, knees pulled up to his chest, sucking on a cigarette. 'I've kind of got much better at being alone,' a woman says to another, as they wander by.

Dark clouds drift across the setting sun, blocking out the light. It feels later than it is. I cut down towards Moray Place, over the road and across the cobbles. In the garden, in the middle of the square, a man walks a whippet. He watches me standing outside the railings watching him. The yellowing leaves from ash trees blow along the gutter and an ambulance rumbles past, blue lights shining on the grand Georgian stone. For a moment, I think the man is going to say something, but he doesn't and locking the gate behind him, he crosses the road and opens a large door – I snatch a glimpse of a tiled hall – then he pulls it shut.

Epilogue

The dog fox

For some time the rabbits went away, and when they came back I decided to leave them be. It had been a strange couple of years and my mother realised on her way back from hospital one night, when snow was falling on the Dalveen Pass, that she couldn't do it anymore, with so few people and a 10-mile drive to pick up a pint of milk since the village shop closed. But it didn't seem real until I got back home and the FOR SALE sign, illuminated by my headlights, was stuck to the wall beneath the cherry tree. Some retiree from the Home Counties had bought our house. He couldn't do it anymore either, so many people and so much noise, but he didn't want our fields, no idea what he'd do with them, so we cut up a little farm that had existed for 100 years. Hugh McLeish, an always-smiling beef and sheep man, bought the land for his son. 'Works too hard, that boy,' everyone always said. 'It's no good working that hard. The boy needs a wife.' I never really knew Mitchell McLeish. He never stopped in his tractor, just one finger raised off the steering wheel, a nod at most, and then away up the lane. Lying in bed, I wondered if he'd come to know the woodcock that fly across the pond, the snipe that live down in the reeds, and the dog fox that lopes along the march dyke as the sun goes down in spring. As the wind whined at my small window, threatening to rip the guttering off again, I wondered if it would ever be the same place for him that it was to me at all.

I slept for a bit and then I woke. The removal men were coming in the morning and all my books had to go into storage. I tried to think of nothing, but instead I found myself thinking of Colin and Katrina and Tom and I wondered whether poetry would be possible in the same way, in years to come, if our birds have gone.

Acknowledgements

Writing a book takes hundreds of lonely hours, but it begins long before that and always involves people who have no idea they were involved at all. I didn't start reading until quite late. At seven or eight, there was always something outside to look at or a fish to try and catch. Eventually, my mother taught me to read, and a love of writing followed. Since then, my father has patiently read almost everything I've committed to paper.

Then there was James Rainy Brown who, in a better world, would still be around to read this book. There was also Paul Williams with the Eliot, Claire McShane with the Durrell, Chispa Prini Garcia who cooked me paella and talked to me about books when I should have been at football practice and Michael Watson who always stayed behind during break to answer my questions on Kesey and Keats.

I wouldn't have written this book if I hadn't spent my first year at university sitting up with my ever-dependable friends, Carlo and Sachin Kureishi, talking about the sort of things we would one day write. If those bleary-eyed nights hadn't been followed by Jenny Batt and Stephen Cheeke's seminars, my interests wouldn't be what they are and this book wouldn't be what it is. A couple of years later, during some otherwise bleak months, Chris Deerin told me I could 'actually write' and in doing so gave me something vital. A short while after that, after I had dropped out of my Masters,

Jonathan Young and Mark Hedges, two of the last great magazine men, made this book possible.

I will forever be indebted to my indefatigable agent, Katie Fulford, who fights my corner at every turn and always seems to be correct in her assurance that everything will be okay in the end. The remarkable David Gothard has been part of this book since it was just a few scribbled notes laid out across the table in a greasy pub on a rainy Tuesday afternoon. He has read every bit of *In Search of One Last Song* and there's hardly a page that hasn't been improved by his strange magic and astute criticism.

My editor, Myles Archibald, is a paragon of patience and has been supportive every time this book has lurched down another hole. A nod must also go to Tom Pickard, who captures the beauty of wild places and the birds that live in them more powerfully than any other contemporary writer and whose poetry sharpened my understanding of sound. Finally, thanks to Robert Vaughan for all his illustrations but especially for those perfect rabbits, all eighty of them, that run through these pages.

Reading and listening

Katrina Porteous says good poetry starts with listening. She's right, I think, and I'm sure the same goes for whatever sort of writing this is too, but it also all starts with reading. I'm not really sure where *In Search of the Last Song* began, but it's been blown about in different directions by a pile of great books that are still currently in storage just north of Dumfries. You should get a copy of Tom Pickard's *Fiend's Fell*, which is full of love for the living wind. You should also try to track down his *Dark Months of May*. These books are in short supply in Britain, where many of our great poets don't receive the recognition they deserve – my copies came from Washington State, with 'no longer property of Tacoma Public Library' scrawled across the title page.

I return again and again to Katrina's *Two Countries*. It is a beautiful collection that chimes with a lot of my thoughts on community, landscape, place and the ever-shifting relationship between them. She has always written her poetry to be read aloud and there are plenty of recordings of her doing it very well. 'An Ill Wind', with its bitter refrain, 'And the knackerman, Aye, the knackerman, The knackerman's in the cattleshed now', captures so many of the ideas that became clear to me when I was on the road, writing this book.

Do listen to Sam Lee's album *Old Wow* on your next long journey. Some of his music is inspired by nature and some of his music isn't his at all but is collected from travellers and gypsies who he's met over the years. I was very moved by Billy Jolly's story about the traveller who was left behind in my chapter 'The souls of our dead'. It is essential that their culture is celebrated.

Alison Brackenbury's collection *Gallop*, including her haunting poem 'Lapwings', which serves as the epigraph for 'The lapwing act', is well worth rooting around for too. She sees beauty, but she also confronts the ugliness of rural Britain, in much the same way as I hope I've done. Enjoy *Gallop* in a Lincolnshire pub. In a similar vein, if you'd like to know more about the poachers Gerald Gray remembers in 'Another church', track down a copy of Lillias Rider Haggard's *I Walked by Night: Being the Life & History of the King of the Norfolk Poachers*. For not much money you can pick up a well-thumbed edition.

I suppose I already knew this, but when I was travelling back and forth and up and down, I was reminded that people are forged by the places in which they live. For an extraordinary portrait of the way that rough country can pickle people, buy a copy of Jonathan C. Slaght's *Owls of the Eastern Ice*, and buy a copy too of John Cumming's *Working the Map: Islanders and a Changing Environment*. I got mine when I went to see John in Orkney to talk about art and kittiwakes, and it made me understand, better than I previously had, that when birds are gone, a place changes, and the people change too.

I never caught up with Colin Simms. Someone I met in Kielder told me I'd been lucky – there'd been some argument about the provenance of a mink pelt in the 1990s and they hadn't spoken since. But, that aside, I think you get a sense of what a remarkable person he is from reading his *Hen Harrier Poems*. It is a collection that made me realise that, if my own book were to be any good, I would have to spend hours and hours waiting, watching and trying to understand.

If you'd like to know more about the doomed Darien scheme, which features at the end of 'Putting down roots', I recommend having a look at Douglas Galbraith's *The Rising Sun*. The scheme marked an extraordinary chapter in Scottish history that fell between burning people as witches and the early Enlightenment. Also extraordinary, in 'Putting down roots', is the life of Thomas Fowell Buxton. The diaries of his gamekeeper, Larry Banville, which were collected and published by HarperCollins, give you a sense of how things have changed, for much better and for worse.

John Burnside's *Giftsongs*, the collection from which my epigraph for 'Putting down roots' comes, can be read over and over without the poems giving away much at all of their wonder and mystery. The same, I think, can be said for looking at birds. We must look, but if we were to ever truly understand, a spell would be broken.

Over the years, I've spent a lot of time sitting in a cold shed reading through the archives of this country's great sporting magazines. They are a rich trove of the wonderful and the grim. Down the decades, different people with different politics have tried to project their own sense of Eden onto rural England's earth. Old men are always dying and their children are always selling their treasured magazine collections for a few pounds. Make yourself sick with James Wenworth Day in *Shooting Times* and make yourself feel better again with Denys Watkins-Pitchford, or BB as he was known, the author of *The Little Grey Men*.

Do support the Norfolk & Norwich Naturalists' Society by buying a copy of James Parry and Jeremy Greenwood's *Emma Turner: A life of Looking at Birds*. It is remarkable to think about how much the Broads she knew, which feature in 'Four long Ukrainians', have evolved. Hers was a world where people cut reed, caught eels and hunted coot. So much is lost so quickly, but it can be dredged up again in books.

When you have a moment, head to the British Library's archive, where you'll find Ralph Vaughan Williams's haunting recording of David Penfold singing about love and the moaning turtle dove. Finally, listen too to Ronny Drew's 'Ratcliffe Highway', which I suppose, in one sense, is where it all began.

Index

Index